THIS BOOK WON'T MAKE YOU RICH

How to have a magically ordinary career with ChatGPT and Artificial Intelligence

By Tim Kulp

ISBN: 978-1-956612-23-3 (ebook)
ISBN: 978-1-956612-25-7 (paperback)

Tim Kulp asserts the moral right to be identified as the author of this work.

Note: This work was produced with the assistance of Artificial Intelligence. Tim Kulp (the human writer) and various Natural Language Generation (NLG) and Text-to-Image (T2I) algorithms, collaborated on this content. The NLG content is produced by an algorithm and then manipulated, rewritten (sometimes), re-imagined, and improved by Tim Kulp. There are cases where what was written by the NLG algorithm was kept in its entirety. Tim Kulp's human mind was the conductor of the creative energies that came from various algorithms and his own creativity to produce this work.

First printing edition 2023
Making Adventure Publishing
16944 York Rd, Suite 63
Monkton, MD 21111

for Evelyn & Victor

CONTENTS

AI STATEMENT

I have used Artificial Intelligence (AI) to assist me in writing this book. AI assisted in brainstorming, writing passages, and editing. I used AI to help me explore, accelerate, and challenge my ideas in this book. AI was a co-writer much like a human collaborator. The various LLMs I used helped me write and brainstorm. No part of this book was taken from AI without review or fact-checking.

If you have questions, please reach out to me at https://timkulp.com/contact.

Enjoy!

Tim

IS THIS FOR YOU?

This book, my dear reader, is for the brave souls who long to embark on an enchanting journey of career growth and transformation without hustling themselves into an early grave.

This book is for you knowledge warriors lurking in your software and spreadsheets. If you find yourself tethered to your computer, yearning to delve deeper into the mystic realm of AI, then this book is your portal. This guide will equip you with the knowledge and insights needed to navigate the ever-changing tides of the AI revolution.

This path is illuminated by the embers of ambition. However, let it be known that this book is not for those who rest on the laurels of "career cruise control." If you have no desire to transcend the boundaries of your current role or ignite a metamorphosis in your work, then heed my words: this book is not for you.

Now, I wish to grant you the power of self-selection. Before you read further, ask yourself: Are you ready to embrace possibilities and scale new heights without sacrificing your fun, family, or friends? This book is for those who seek balance, who desire growth without surrendering their lives to the relentless pursuit of success.

Now, dear reader, as you stand at this crossroad, it is time to choose your path. Will you let others control your career destiny in the Age of AI, or will you step into the light of knowledge? If you have made the choice to embrace your inner adventurer, then grasp this book tightly, for within its pages lie the secrets to navigating the labyrinthine world of AI.

Let us embark on this journey together, my friend, and unlock the hidden potentials that await us in the Age of AI.

INTRODUCTION

Why would I write a book that can't make you rich?

For most of us, AI isn't going to make us rich. With that said, I'm guessing you'll need to work tomorrow. And the day after that. And, if you're reading this book, probably the next few years too. AI will change work, and I wrote this to help you navigate that change while keeping your sanity.

Every time I open YouTube, I'm bombarded with videos about how to make $10,000 a day with ChatGPT or 25 AI Side Hustles. If that's for you, go for it, but I like my free time. I love my job; I love my coworkers. I love my family more and don't want to be hustling 24/7. If you're like me and want to enjoy work, explore how AI is changing your job, but keep ahead of it and stay relevant to your business, this book is for you. If you're hoping for some get-rich-quick scheme, look elsewhere. AI is not a magic money machine, but it can be magical in a different way.

Imagine a future where you only did the parts of your job you loved. Or a future where your team was the creative force driving a digital army of workers that did all the grunt work. A future where you are empowered by AI instead of replaced by it.

We are early in the Age of AI and the decisions we make today on how to work with AI will drive the future. This book is a guide on how to create a future that doesn't suck, one job at a time. While tech companies fight for AI supremacy, we workers need to craft a future where AI is part of the equation and not the end of the line.

I believe the future of work is bright if we put the effort in to make it shine for everyone.

Your career is no longer a solo sport, it's a team sport. While many of us were raised in the "pull yourself up by your bootstraps" myth, modern office work is impossible without collaborating across teams. Win as a team or lose as a team. The future of work is not human vs. machines but human + machine in a seamless dance of digital, human and robot. A new player might be ready to join your team: AI.

WHY ME?

I struggled with whether to write an AI book. Do we *really* need another one? But then I noticed a trend in the content being produced about AI. It's all about change. Where are the books for people who are just doing their job and co-existing with AI at work?

I'm providing a counterpoint to the "everything's going to change" and "prepare for the tsunami of technology" texts that are interesting and helpful for business, but not written for the worker. This is for the worker.

Professionally, I've been helping organizations with their

digital transformation for over 20 years. As a mentor, I've been helping people transform their career for the last 15 years. As a father, I have two children who will enter the workforce long after this AI revolution has taken root and grown. In my community, I see both the potential and harms of algorithmic tools in everyday life.

I am the Chief Innovation Officer at Mind Over Machines and have been helping clients, from small non-profits to global enterprises, grow automation and AI capabilities. Like people, companies have challenges with change and digital transformation. No one jumps into the deep end of a new technology without slipping under the water at least once.

Since my early developer days, I have been helping businesses through their digital transformation. In all those projects, I found the key to the future of work isn't in training or features, it's in building digital confidence in the users. Deploying new technology doesn't help if no one uses it or has the confidence to explore the value it promised to provide.

Mentoring is one way to build that confidence, as is teaching. For the past four years, I've been working with community colleges, high schools and universities to grow AI skills in the next generation workforce. Together with these institutions, I've built a few AI curriculums, taught many classes to students and business, and forged apprenticeships that took students beyond theory to practice.

This goes beyond building an army of data scientists. We need an army of AI Literate people who can see AI's use cases in their everyday job. If business implement AI tools without the staff that understands them, how to use them, how to drive value from them, that investment in AI was a waste. AI enables people but only if the people know how to use the tools.

AI is perceived to be the domain of technologists. I am here to change that. We need everyone to participate if we want a future where AI empowers humans. Consider this my invitation to you to begin your AI Journey. I've helped many others get started and now it's your turn.

WHAT ARE YOU GETTING OUT OF THIS BOOK?

I want you to have actionable steps to grow your AI knowledge and apply it to your day job. This book isn't about side hustles, climbing the corporate ladder, or hustling yourself to an early grave. It's about enjoying life while exploring a technology shift and engaging with your peers to build the future we want.

By the end of this book, you will know how AI tools could apply to your work. This isn't a theory book. It is a workbook. Exercises throughout the book will encourage you to apply what you're learning. Every chapter ends with a meeting to schedule for yourself to act on the content so you can invest in your growth.

My guideposts for this book were: Actionable, Realistic, and Inspiring. I'm hoping you'll find this book all three and start your AI Journey with confidence.

I'm excited about what we can build together.

Let's get started!

Tim

CHAPTER 1: YOUR MAGIC MAKES AI MAGICAL

There is a paradigm of thinking focused on YOU needing to meet AI as your new coworker. Let's flip that to start. The AI system needs to meet YOU!

Here's the reality: the AI system probably wasn't built for YOU. It was built to do a purpose and you are the user of the system. It doesn't care if you are you or someone else. The AI system is indifferent to the world and performs its function regardless of the user.

Now, for the wonderful news, ***you*** are the key to converting that indifference from function to magic.

And when I say magic, I mean the spark that no one else has. I'm not talking about you being a unique gem, your history, your experiences, I'm talking about how all the things that make you YOU combine to create an ability that is unique.

In business, we refer to this as a **Core Competency**.

THINK ROOTS, NOT LEAVES

In The Core Competency of a Business by C.K. Prahalad and Gary Hamel (HBR 68 (3)), the authors describe a business like a tree. The leaves are the product that is delivered to customers. Driving the branches and trunk that yields the leaves is the root system. These roots are the Core Competency, and they are the capabilities that will separate you from everyone else in the Age of AI.

Crash Course in Core Competency

Want to dig in deeper on Core Competency? Harvard Business Review (HBR) has an excellent explainer video on the topic that illustrates the concepts discussed here. https://hbr.org/video/5146717725001/the-explainer-core-competence

Often, we compare ourselves based on product and regard ourselves as interchangeable in our work. A person who is a medical claims processor might see themselves as the same as all the other medical claims processors, and this is a terrible situation. Who wants to just be a cog in the machine of work? Who wants to work in a role where no one cares if *they* are doing the work or someone else is? If you are like everyone else, you are a commodity, which means the work you are doing can be replicated by a cheaper resource with no loss in value. This is where AI takes jobs.

You are more than the output of your work. Your amazing roots create unique leaves. But no one is sitting around searching for how your leaves differ from anyone else's. You must show them how your roots create valuable leaves.

EXERCISE: IDENTIFY YOUR CORE COMPETENCY.

Spend the next week reflecting on your career and list out what you think are your strengths and weaknesses at work. Then talk to colleagues who have worked with you in the past, ask them for their honest opinion about which of those qualities they feel make you stand out as a unique individual contributor. Take notes from this conversation to form a comprehensive picture of how your core competencies helped shape your success in previous roles or projects.

In future exercises, we'll explore how AI can help. Your Core Competencies will be key throughout the rest of this book, so post them somewhere clear. I keep mine on a sticky note at my desk. Other people I've worked with hung their list on the wall in a picture frame. Make it visible so you can always be reminded of the value you bring.

Here are some sample Core Competencies from people in my classes and workshops:

- Curious mind driven by a liberal art background.
- Commitment to constant learning and growth.
- Deep passion for Python code and mastering my craft of software development.
- Unique ability to hear both sides of an issue and help people navigate the conflict with humility and a relationship focus.

Notice, these aren't "no one else in the world can do this" but they are unique to the person and within teams of people. You are not trying to be unique in the world, you are trying to illustrate your unique value within your business.

YOUR JOB IS NOT "A" JOB

We think about our work as an atomic thing (atomic as in a single, inseparable thing not explosive...hopefully). Our job as a Project Manager is to do the work of a Project Manager. Our job as an electrician is to do the work of an electrician. And our job as a manager is to do nothing and boss other people around... just kidding managers. Here's the truth: none of these jobs, nor any job, is a single thing. All jobs are composed of tasks that must be accomplished.

EXAMPLE JOB: EVENT PLANNER

Task 1: Research potential venues and determine their suitability for the event. This includes evaluating space availability, catering services, design considerations and overall cost.

Task 2: Create a budget for the event that outlines all necessary expenses, such as venue rental fees, catering, decorations, entertainment and other costs.

Task 3: Plan logistics, such as transportation, accommodations, catering services, entertainment and any special equipment needed for the event.

Task 4: Develop a timeline for the event that outlines when tasks need to be completed, such as when deposits must be paid and how the event will be promoted.

Task 5: Coordinate all vendors and ensure that each one is meeting their obligations for the event. This includes arranging for necessary permits, insurance documents and other paperwork.

Task 6: Monitor the event to ensure all vendors are meeting their objectives and that the event is running smoothly.

We'll stop, but you get the point. Even each of these tasks can be further broken down into subtasks.

Task 1: Research Potential Venues and Determine Their Suitability for the Event

Subtask 1: Create a list of potential venues.

Subtask 2: Research each venue to determine suitability for the event.

Subtask 3: Evaluate space availability.

We could go on for a long list of subtasks for each task. The subtask should be atomic with a single outcome. Beware lumping subtasks together with different outcomes, such as evaluate space availability and design considerations for the space. The "and" tells us this is two subtasks.

Jobs are made up of tasks. Tasks comprise subtasks and so on. You get the idea at this point that your job is not a single thing.

This is where AI comes in. Some of these subtasks are better done by a human, others are better done by a machine. But how can you tell which is which?

WHAT IS YOUR CORE COMPETENCY?

To answer what tasks you should do and what AI should do, start with your Core Competencies from the previous exercise. If a task hits your Core Competency, then you are bringing some kind of magic to that task.

As an example, one of the core competencies shared by Harvard Business Review is Southwest Airlines' focus on providing cost efficient flights. By building this core competency and implementing it everywhere in their business, Southwest became a leader in low cost flights leading to market domination in US domestic flights. Could others have low cost flights? Yes, but because Southwest focused on it and made it a core of their business, they were able to infuse everything in their business with that magic before the competitors could catch up.

But having magic isn't enough. Is your magic worth the cost of your time? If your magic is the ability to imagine the light in an event venue at a certain time of day which enables you to identify the perfect moment for a wedding, that is amazing. But that doesn't mean you should tour every wedding location before deciding on any location. No, you would use technology to help you shorten the list to a few options you tour. If you spent all your time touring every potential wedding location in New York or even a smaller city like Baltimore, you wouldn't have time for other parts of your business, like meeting clients.

As technology allows us to do more, your time remains your most precious thing. How can you use your time to best effect, so that you are using your own Core Competency and AI to help you become the best event planner?

By using AI in different ways, such as automating processes or data gathering, it allows you to focus more of your energy on tasks that require human touch. This could be the ability to decide without complete information or understanding the nuances of customer service. As an event planner, you must navigate the complex emotions your clients have regarding the event. AI can enable more time for dealing with emergency melt downs while not missing a beat on other aspects of the event.

You have been able to use technology to make your job easier for some time. Spreadsheets make tracking and analyzing data easier than doing it on paper. Calculators help with math. Pencils are easier to use than burning your own wooden sticks. AI is now joining the list of technology to help you work. It is here and it can help you bring your Core Competency to every task you do.

WHAT IS AI'S CORE COMPETENCY?

So, when should you use AI? Start by recognizing AI is not just one thing. There are many types of AI systems, and they are often specialized to a task.

The Netflix movie recommendation system is a great example of a hyper specialized AI. It gives you recommendations on what to watch next but does not schedule alarms on your phone for when you should watch them. This is known as Narrow AI: an AI system that is very specialized for a single task. Other examples would determine whether to approve someone's insurance claim, reading an invoice with computer vision, or predicting the likelihood of a car accident at a specific intersection based on traffic patterns. These systems only do what they've learned to do.

Beyond Narrow AI is the concept of Artificial General Intelligence (AGI). This is where the AI system can learn on its own and do new tasks that it was never taught. We're not here yet. For now, be aware there are many people working on building AGI, but we're far from it being a reality for now.

Between Narrow AI and AGI, is a type of AI that is all the rage since ChatGPT hit the world stage called Large Language Models (LLMs). These are often used as text completion or content creation systems, but that is a significant miss on their true potential. AI luminary Andrew Ng refers to LLMs as Reasoning Machines in his LLM Prompt Engineering course (https://www.deeplearning.ai/short-courses/) and he's right. These systems are not mere answer engines like a Search Engine, they can do so much more.

These LLM Reasoning Engines are a General Use tool. They do not have a specific purpose but can have many purposes. They can discover information. Write information. Explain ideas. Generate ideas. You can question them, engage them.

We will dig deeper into the concepts of LLMs in a later chapter, but for now, let's examine a task from our above list that an LLM can help us perform.

Subtask 1: Create a list of venues: This is a perfect LLM task where you could ask the LLM to create a list of venues in a specific location for you to consider for a wedding. Here's a sample prompt:

"Act as a wedding planner and provide a comprehensive list of wedding venues in the Mount Washington area of Baltimore City that are suitable for a spring outdoor wedding. Your list should be ranked in order of preference and include further details such as capacity, amenities, and availability. When researching the venues, consider important factors such as the view, accessibility and surroundings, keeping in mind that the couple is looking for a serene environment that captures the beauty of spring. Summarize each venue, covering their key attributes and unique features that set them apart from other venues in the area. Be sure to keep in mind the couple's preferences for their wedding, including any specific requests or themes they have mentioned to you. Ideally, your list should include both budget-friendly and high-end options to suit all types of wedding requirements."

With this prompt, we get the following output from the LLM:

Based on the information gathered, here are some of the top wedding venues in the Mount Washington area of Baltimore City suitable for a spring outdoor wedding:

Mt. Washington Mill Dye House: This venue offers a unique blend of history and modernity, perfect for couples looking for something different. It provides multiple spaces and flexible floor plans, allowing you to tailor your special day to your liking. The outdoor space is perfect for a spring wedding, with beautiful views of the surrounding area. Source: Dyehouse Events

Gatherings By Woodberry Kitchen: Located in the heart of Mount Washington, this venue offers a distinct charm with its rustic yet elegant setting. The outdoor space is great for an intimate spring wedding, surrounded by lush greenery. Source: Yelp

Gramercy Mansion: This venue offers a serene environment that captures the beauty of spring. With its beautifully manicured gardens, this venue provides a romantic setting for your special day. Source: Yelp

Waterfront Wedding Venue: Located on Baltimore's beautiful waterfront, this venue includes both indoor and outdoor event spaces with stunning views. The surroundings offer a serene environment perfect for a spring wedding. Source: Wedding Rule

Main Street Ballroom: Although not located directly in Mount Washington, it's a short drive away and offers a versatile indoor and outdoor space that can be transformed to fit any theme. Source: Justin Kunimoto Photography

Remember, each venue has its own unique features and amenities, so it's best to visit them personally or do a virtual tour to get a feel for the place. Also, availability can vary, especially during the peak spring wedding season, so it's advisable to contact these venues as soon as possible to secure your preferred date.

Now, with that output, we can use the LLM to help us research each of the venues for more information. Once we've narrowed our list down, we can see the best candidate venues.

Warning!

Two things to remember when working with LLMs: they can make things up and they can be wrong. I'll talk about this at length in the LLM section, but for now, be aware that LLMs are not always right.

A practice I encourage you to adopt is what I call checking the "diamond in the rough" (a reference to Aladdin). Your AI system will give you recommendations, but you are not limited to what your AI suggests. Find something it did not recommend and check it against the recommendations. In your HR recruiting system, select a few candidates that were automatically disqualified and see why they were disqualified and whether you agree. In our venue example, find venues not on the list and check them out. Should they be on the list? These systems are not all knowing and they can easily miss a diamond in the rough.

Finally, remember, you are driving the work. The AI is your sidekick, not your navigator, and you decide the path to take, the items to investigate, the recommendations to pursue. Own your work.

EXERCISE: TASKIFICATION

Taskification is a creative and structured way to break down a complex process into smaller and more manageable parts. This exercise should help you get started.

Step 1: Think about the things you do every day for your job.

Make a list of these things. This list is your core task list. Define a task as something that creates a specific outcome. Use a bullet list in your word processor (Microsoft Word, Google Docs) to list your tasks.

Step 2: For each task, identify the primary goal of that task. Why do you do that task? If the answer is, I have to do it because my boss said so, then write that down. Some tasks you'll have no idea why you do them. They fall into the "we always did it that way" bucket. At work, we often go on auto-pilot and just do the work without thinking. This is your moment to consider what you do and why. If you do not know why you do something, talk with your peers to see if they know. Create a sub-bullet under your task that lists the goal like this:

- Task 1: The thing you are doing
    - Goal: Why you are doing it

Step 3: Then, break it down into its key elements, or subtasks. For each subtask, evaluate the resources you need to achieve it and identify what steps are necessary to complete it. Make these a list under each task.

- Task 1: The thing you are doing
    - Goal: Why you are doing it
    - Sub Task 1: The first step to perform the task.
    - Sub Task 2: The second step to perform.
    - Sub Task n: The rest of the steps.

Step 4: Finally, create a timeline for completing each step of the task. Does a subtask step take 5 minutes or 5 hours? Write that at the end of the subtask. You don't have to be perfect or exact. The goal of this exercise is to map what you do and determine how long each subtask takes you. This information will help you identify automation opportunities.

When I lead a Taskification exercise, I draw it like a tree. The job is the top box, then I draw a bunch of connected boxes that are the primary tasks for the job. Under each task, I draw another series of subtasks that make up the primary task.

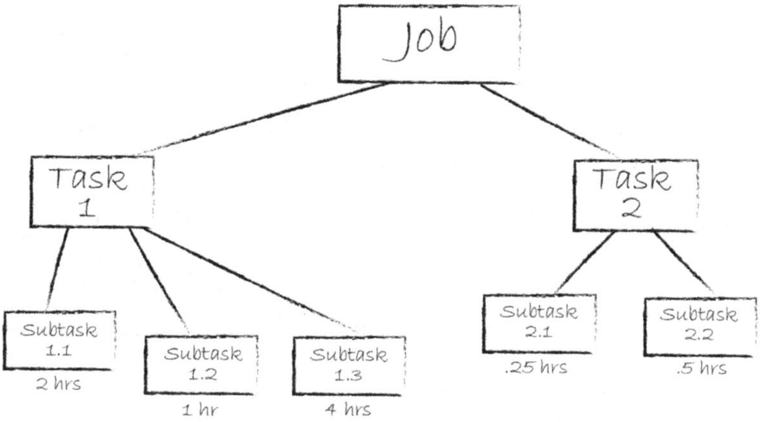

This format helps me understand all the items and how they are related to each other. I use numbers in the Subtasks to illustrate their order. Stop at the subtask level for now. Later, you can go as deep as you want, but for this exercise, focus on Job > Tasks > Subtasks.

When you're done, ask yourself which ones could be done by an AI coworker. Which ones can be automated? If you're unsure, keep reading. We'll explore use cases and sample automations as we continue.

Share your task tree with a peer and ask if you missed anything. Have fun.

SKILLS ARE THE FUEL FOR TASK COMPLETION

You are a skilled worker. Those skills are used to complete your tasks.

As technology continues to advance, having the right skills in a business is becoming increasingly more important for employers, even in traditionally non-technical roles. As AI and automation become more prevalent, it is essential that employers focus on hiring people with specific skills that are valuable in an increasingly competitive job market.

LinkedIn's Global Talent Trends 2019 report found that soft skills are the most important factor in hiring decision. These skills, such as creativity and adaptability, help employees remain competitive even with changing technologies.

The great news is that skills can be learned. Employers should create an environment in which their employees are encouraged and supported to develop new skills. Learning platforms such as LinkedIn Learning can help you gain the skills you need while still working your day job.

But you will probably be on your own. The good news is that we develop skills while we work and can learn new skills when we approach learning intentionally. The even better news is that you are building skills every day you go into work. Every thing you do is building skills if you are intentional.

Intentionality is the name of the game. As an old student of mine once said, "Practice doesn't make perfect. Perfect practice makes perfect." I love that statement. Part of Perfect Practice is knowing what you are practicing, why, and how you are trying to grow from the activity.

Let's look at another job, the tasks, and now the skills.

JOB: REAL ESTATE AGENT

1. Research properties that fit customer criteria

2. Visit potential homes with customers

3. Negotiate offers on behalf of clients

4. Analyze market trends and property values

5. Assist customers in obtaining financing for their purchase

6. Prepare legal documents such as contracts, deeds, leases, or other agreements

7. Monitor the housing market to remain informed about pricing and regulations

8. Attend open house events to find buyers or sellers

9. Manage financial aspects of transactions, such as down payments and closing costs

10. Develop relationships with lenders, appraisers, home inspectors and other real estate-related professionals.

Now, let's dig into Task #2, what are the skills needed for Visiting Potential Homes with Customers.

SKILLS FOR TASK #2

1. Communication skills to listen and articulate customer needs
2. Research skills to build knowledge of local market trends
3. Patience and a friendly attitude (yes, these are skills)
4. Attention to detail
5. Organizational skills
6. Time management skills

These skills do not enable just one task, but many. Notice the relationship of the skills across tasks. Let's take another task and explore it. Task 3, negotiate offers on behalf of customers.

SKILLS FOR TASK #3

1. Negotiating skills
2. Communication skills
3. Knowledge of legal documents and regulations
4. Problem-solving
5. Research
6. Analytical thinking skills
7. Persuasion/influencing skills
8. Patience and diplomacy
9. Integrity and trustworthiness
10. Time management

EXERCISE: SKILLS INVENTORY

Now, let's figure out what skills you have. This is where we are going to build on top of our task list and subtask list.

Step 1: Review the tasks that you perform regularly at work. Make a list of all the skills needed for each task or subtask. These can be hard or soft skills — anything from coding expertise to communication ability. Think about which areas you already excel in, and which ones could use improvement based on how frequently they come up in your daily job activities. Divide these into two categories—"must have" and "nice to have" skills - so that you can prioritize them according to importance/difficulty level and relevance based on current trends in your field.

Step 2: You are biased. You might not know the skills you need for a task. Ask your digital coworker, your AI sidekick, the following question:

"You are a recruiter for [company name]. You are hiring for the role of [your job title]. That job performs the following tasks [list your tasks]. What skills are needed for these tasks?"

Use ChatGPT, Llama, Jasper, whatever the right LLM is for you and see what you get for the skills list. Then, gut check yourself on which of these skills you have, which you should have, and which the LLM totally missed.

Step 3: Share this inventory with a peer or mentor who may offer additional insights on areas where development will benefit performance overall; peer feedback is invaluable!

This exercise leaves out an important skill set from your inventory. These are skills that have nothing to do with your work. Are you a superb Dungeon Master in Dungeons & Dragons? Are you an encyclopedia of horror movies? Can you coach 30 8-year-olds and get them all moving in the same direction? These skills, while not directly related to your day job, can be incredibly valuable at work. Dungeon Masters are excellent communicators. Coaches can lead others. Movie buffs have excellent knowledge retention and recall. Bring your entire self to your skills inventory and Core Competency. Remember, this is your magic and you shouldn't hide it.

Try it at home

Work isn't the only place where these skills will help. You can use exercises like Taskification in all aspects of life. Try it out. Explore building a list of tasks and subtasks for the following job: Making a Pizza.

What are the tasks? Subtasks? How long does each subtask take? Map it out and see what you find.

THE ALCHEMY OF YOU

You have your tasks. You understand what you should do and what AI could help with. You have a list of your skills. You have a list of your core competencies.

Aligning core competencies, tasks, and skills is essential for individual contributors to be successful in their role. It allows you to focus on what you do best and leverage technology to automate the mundane or repetitive tasks that don't require human input.

This will help you stay competitive in a rapidly changing job market as well as maximize productivity by allowing you to focus on higher-level activities. By being aware of your strengths and weaknesses, you can better assess where AI can help you become more effective and efficient.

Now, put it all together with this reflection exercise.

REFLECTION

I like to use notebooks to capture my exercises and journal my thoughts. Using your own journal, write your answers to the following questions. Date the page so you can come back in the future and read what you were thinking at this moment.

- How does your core competency appear while executing your tasks?
- Do your skills align with your core competency?
- Which skills are used most often in your tasks?
- Which of your tasks do you love to do?
- Which of your tasks would you prefer to hand off to an AI coworker?

BUT WHAT ABOUT AI?

You have all this magic, the Alchemy of You, coming together. AI is an ingredient in that alchemy to scale your magic. AI can enhance your skills or take on tasks that you don't enjoy doing. AI can help automate and streamline certain processes within your job role so you can focus on the tasks that really use your core competency.

AI is not meant to replace humans; it's meant to augment our abilities and create new opportunities for growth. So, how can you use AI to enhance your core competency? This is the question that will bring us full circle.

Not that your magic will always save your job. If there are financial pressures that outweigh human value, AI could replace you no matter how magical you are. This is a cultural decision by business, not a fore-gone conclusion. If your business values people, values your magic, AI will not replace you if you communicate your value and magic effectively, widely, and repeatedly.

AI can be a powerful supplement to any resume. Don't overlook what you think is common or what everyone else is doing. Highlight your use of AI in your everyday work by listing it as an accomplishment on your resume. Remember to show how you made the AI more effective. Combine your use of AI with your magic and you'll differentiate yourself from everyone else.

Never underestimate the power of differentiation, nor the value of showing a successful track record working with AI.

WHAT To DO ON TUESDAY?

I'm assuming you are reading this book during the evenings or on the weekends. You are building some ideas here, but let's be real. We all have day jobs and it's hard to find time for anything.

Each chapter will end with this section "What to do on Tuesday". I want you to schedule a meeting with yourself on the following Tuesday of whenever you read that chapter. Book the time specified in this section with the agenda provided.

My goal in these sections is for you to create time to reflect, time to engage with yourself, and time to explore this new world we're building together. There is nothing magical about Tuesdays. You can select whatever day works best for you. I always think of Monday as my "prepare for the week" day, Thursday, and Friday are my "wrap up the week" days. Wednesday is the mid-point and often a check-in on the week's progress, but Tuesday is just there. So, I make Tuesday my catch up day where I focus on staying in tune with what's important for my job and not just getting tasks done. Select whatever day works for you.

Also, I am using the 7Ps framework designed by James Macanufo. I discovered it through the book Gamestorming by Dave Gray, Sunni Brown and James Macanufo. Learn more about the 7Ps on the Gamestorming website at (https://gamestorming.com/).

Timeblock: 30 minutes

Purpose: To examine my core competencies, tasks and skills

People: You

Process:

- Setup your space to be distraction free (this is critical)
- Complete the Core Competency exercise with at least 3 core competencies

- Complete the Task list exercise with at least 10 tasks and subtasks listed out (10 items in total).
- Complete the Skills exercise with at least 10 overall skills across all tasks (don't forget to use your LLM sidekick when necessary)
- Schedule a session with a peer to share your lists with them and get their feedback (perhaps treat them to coffee or tea)

Product: The Core Competency List, The Task Tree, The Skills Inventory, and one meeting with a peer to share what you've built.

Pitfalls: If you do not share, you will not create the feedback loop that is critical. It is important that you engage a trusted peer to share what you're thinking and ask what you are missing. Also, 30 minutes is not that long. Don't spin on these things. Leverage your AI assistants when necessary and keep to only 5-7 minutes per exercise. You can always schedule more time to refine and iterate through the exercises.

Prep: Set up a distraction free environment for yourself so you can go all in on the activity.

Practical Concerns: The blank page can be scary. Don't get tied up in being perfect. Write the first thing that comes to your mind and remember, you are not carving this into stone. You can erase, evolve, iterate and re-imagine as needed.

Enjoy and talk to others about what you're doing. The social collaboration is so powerful. It will help you engage and connect with others and be clear about the magic you bring to work. Remember, AI can't bring YOUR magic to the work. Be proud of your magic. Tell everyone and celebrate your magic, their magic, everyone's magic.

CASE STUDY: MAGICALLY UNAWARE

I once worked with a wonderful team member, we'll call them Lewis, who was fresh from university and wasn't sure where his career was going. Lewis was a generalist in the organization and he was the personification of "other duties as assigned" doing a bit of everything.

As a generalist, this person sat in on a design session as the note-taker where we were brainstorming on a new user interface. Lewis sat quietly while we sketched and doodled on the whiteboard. When we engaged him to participate, he stayed quiet.

After the session, I asked if Lewis felt comfortable in the discussion and ensured him that his input was not only valuable, but very much wanted. He explained he did not participate because he was too enthralled with how the discussion was going and didn't want to mess up the flow.

During the next session, I invited Lewis to join again and this time he was so excited to join. It filled all of us with a whole new level of energy. This kept happening with each new session.

Lewis had never heard of a User Experience (UX) Designer, but when he saw the work in action, it ignited passion within him. He sought a mentor to help build UX Design skills and planned to exit his generalist role for a new UX role, but there wasn't a UX role in the company.

He was still a generalist in the organization but pursued UX Design tasks by telling everyone that was his interest. Because of that, he was considered and chosen to take part in UX Design sessions. Eventually, he grew the skills to perform UX Design on his own. These skills were built both on the job and at home through personal projects.

Lewis applied for and earned a Jr. UX Designer role at another company with the skills he built doing the tasks he volunteered for in his old position.

I talked with Lewis for quite a while after he started his UX Design career and while we never discussed his core competency directly; it was the ability to absorb information, file questions away in his mind for exploration later, and his curiosity that helped him forge his UX career.

Last I checked in on Lewis, he was leading UX Design teams, focusing on mentoring and helping others find their passion for intuitive design.

CHAPTER 2: YOUR SIDEKICK

When you start a new job, you need to meet new people. Few jobs are isolation booths where the only person you interact with is yourself. For modern jobs, Artificial Intelligence (AI) could be one of those new people. In this chapter, we'll get a better understanding of our new AI sidekick.

But let's get something out of the way immediately. AI is not human. I know that's obvious and might sound a little silly, but people commonly imagine their digital coworkers as no different from their other coworkers. I've worked with many organizations who quickly name and personify their AI systems. In team meetings, new people wonder who Robi is and even ask if they're remote or in the office. Robi was an RPA bot (more on those later). As humans, we are eager to anthropomorphize and this is a hurdle that every business, every team, everyone needs to navigate.

Anthropomorphizing is when we think of things that are not human as if they were human. For example, when you give your toy robot a name and assume it has feelings, this is anthropomorphization. Just like giving your pet dog a name and imagining Rover can do things that only humans can do (like be passive aggressive). Humans love to see ourselves in other things. This isn't a bad thing by default, but always remember, your AI tools are machines. They are fallible just like any other machine, no matter how human they seem.

AI IN THE WILD

There are lots of ways Artificial Intelligence can show up at work, but before we get into that, let's get a common definition on how to recognize AI. Unfortunately, the words Artificial and Intelligence are difficult to explain individually, much less together.

Here's how my AI co-writer explained AI:

> *Artificial Intelligence (AI) is a branch of computer science that studies the processes and methods used to create machines that can think and solve complex problems similar to humans. AI is an interdisciplinary field of study that combines multiple disciplines, including computer science, mathematics, psychology, neuroscience, linguistics, philosophy, and engineering. It has been around for over sixty years, but in the past decade AI has leapt into the public consciousness and is becoming increasingly common.*

And here's what Merriam-Webster has to say (https://www.merriam-webster.com/dictionary/artificial%20intelligence):

A branch of computer science dealing with the simulation of intelligent behavior in computers

The capability of a machine to imitate intelligent human behavior

But using these definitions is like saying the following when talking about a dog (this is from my AI co-writer when asked "what is a dog"):

A canine is an animal of the family Canidae, which comprises domesticated and wild dog species, wolves, foxes, jackals, coyotes, and dingoes. They are typically four-footed mammals that possess large teeth and long legs for running. Canines have a sharp sense of hearing and smell combined with a superior night vision.

This is a useful definition of the canine family, but if you see a four footed animal with large teeth and long legs in the wild, would you think *oh, that's a canine*? No, you would compare what's in front of you with dogs you've seen before to determine if the approaching animal is a dog, a wolf or a chupacabra. You would not mistake a dog for a wolf, or a dog for a vampiric livestock devouring monster. Nor would you consider these animals similar.

AI works the same way. There are different animals in the AI kingdom and knowing which does what is important to knowing when to apply an AI animal to a particular task.

Let's create a shared definition of Artificial Intelligence for ourselves in this book. This definition is my own making pulled from many great insights from others. While not fully comprehensive, this definition gets us started with a simple view of AI that we can recognize.

AI is a software system that learns, observes, and reacts to its context/environment.

This definition breaks down into a few parts:

- **Software system**: code that runs on a computer and not in a biological system
- **Learns**: the software takes data from pre-existing datasets to decide on how to react
- **Observes**: the system uses its present context/environment to determine its response in real-time.
- **React**: the AI does something more than think

There are many other conceptions of AI that you can explore. IBM has a wonderful webpage that digs into a lot of AI concepts and academic definitions (https://www.ibm.com/topics/artificial-intelligence). Explore the various definitions of AI to expand your knowledge of how others are approaching this field. My goal in our shared definition is for you to have a mental checklist when you look at a software tool to ask yourself, is this AI?

Let's explore sample AI tools we will find in the wild, starting with a common, useful one that can help with your tasks.

ROBOTIC PROCESS AUTOMATION (RPA)

A few years ago, RPA was all the rage. Companies like Blue

Prism, UiPath, Automation Anywhere and Microsoft were creating tools that help businesses do amazing things with their old software systems. Since ChatGPT came along, RPA isn't getting all the spotlight it once did, but this AI tool is extremely useful and can help supplement your tasks immediately...at least that's what the marketing content says.

Robotic Process Automation (RPA) is a form of Artificial Intelligence (AI) that automates repetitive, rule-based tasks. It helps businesses save time and money by taking over manual and tedious tasks from everyday workflows. RPA allows you to create, execute, and monitor tasks without manually coding them. This allows you to do more with less and streamline processes.

RPA tools parse data from different sources, trigger relevant actions (such as sending emails or creating Helpdesk tickets), and provide insights about your process through analytics. For example, a business may use an RPA tool to collect customer feedback data from emails and store that data in the Customer Relationship Management (CRM) system. This allows you to have a central place (the CRM) to see all the interactions with your customers and understand why Customer X is so frustrated while Customer Y is elated.

The secret power of RPA is in two aspects, the Process and the Automation. Often when people do work, we don't think about what we're doing. We just get things done. Consider how difficult writing a step-by-step guide of how to do your tasks would be.

A core benefit I've seen with RPA is the deep dive into understanding the business process you are trying to automate. Often, people don't know how they perform a process and as they go through the exercise of documenting the process, they see opportunities for efficiency before any automation is applied. When working on any RPA project, you must write the process down to the click (as in, every time you type something or click on something on your computer). This is important because this level of detail is what will inform the developers on how to build the automation. Think of this documentation as the instruction manual on how to perform a business process.

SIDEBAR: What is a business process

A business process is the sequence of tasks that a company carries out to create value for its customers. It is a set of linked tasks that use resources from people, systems and machines to transform inputs into outputs. A business process typically has a start point, an end point and multiple steps in between. At each stage, specific decisions are made based on the input received.

Once you've built your process map, you can see which tasks can be automated. Which tasks can be automated depends on a few factors, the key one being which tasks do not have value added by a human. If a task is completed just as well by a system as by a human, then have the system do it. Tasks that are done by humans should have some level of human magic brought to them.

But can the RPA tool you have chosen do the task? The capabilities of RPA tools are similar, but like all software, different vendors have focused on different things to make their RPA tools unique. Microsoft focused on making their tools easy for non-developers. UiPath focused on simplifying the ability to find objects on the screen (called Selectors). Others have their own focus. The marketing material for many RPA products states these tools can do many things, but when you experiment with them, you'll find that the marketing material is often wearing rose-colored glasses.

These automations can do simple things like entering text into a textbox on the screen, pressing buttons, copy and paste data from one application to another. They can also do more complex things, such as reading invoice forms and transcribing the data within them into your finance system. The best way to discover what your RPA tools can do is to experiment with them. There are tons of videos on YouTube for all these tools demonstrating their features and use cases. Many of them offer community editions and free trials for you to explore.

RPA vendors will tell you their tools can do anything. Make sure you vet your use case thoroughly. These tools are not magic, they are just software and they can't do everything. But, what they can do, they can do very, very well. Explore them to understand where they fit into your task list.

WHEN To USE RPA:

Use RPA when you have a task that is low-variance (doesn't change much), high volume (happens a lot) and low-human (you don't bring much magic to the task). RPA allows you to take the busy work out of your task list so you can focus on the higher value human work.

SAMPLE RPA USE CASES:

- Copy and paste data between systems (often called Swivel Chair Operations)
- Processing data with simple, definable reasoning
- Reading content and acting on that content in a business process (such as reading emails from clients and escalating those emails when the client becomes agitated)
- Performing the same action on a queue of items

There are many more use cases, but these will help you think about tasks that you have and how RPA might fit.

EXERCISE: PROCESS EXPLORATION

Now we will build out a process map to help you identify the RPA opportunities in your workflow. But before we begin, we're now at the point of this book where you will explore AI. Before you do anything in production, that means using real data in your business, communicate and collaborate with your leadership on what you are doing and why. Chapter 3 dives deep into preparing your business for AI. Consider all this here experimentation and never use business data from actual clients, users, patients, or any other real thing.

Ask your IT team if there is a Test system to try your new AI skills on. Share with them what you're doing and how they can be a great partner for you. Seek collaboration, not obfuscation.

Step 1: Identify the Process

Choose a specific business process that you want to map out. Processes related to customer onboarding, inventory management, are great places to start but use any process that is important in your job. Select an extremely simple process.

Step 2: Define the Start and End Points

Identify where the process starts and ends. What triggers the process to begin, and what is the outcome or goal of the process? For example, customer onboarding is triggered by a new customer contract in sales, or inventory management is triggered by a restocking truck.

Step 3: List the Steps

Break down the process into individual steps. Start from the beginning and list each action or task that needs to be completed to move the process forward. Be as detailed as possible, including any decision points or branches.

Step 4: Determine Inputs and Outputs

For each step, identify the inputs (information, documents or resources) required to complete the step, as well as the outputs (results, reports or actions) generated by that step. This information is important to understand what data drives the process actions. Sometimes the drivers are simple logic in your head. AI can't help you if you "just know" what to do. Document what you "just know" so the AI can help you.

Step 5: Map the Flow

Visualize the process flow using a process mapping tool or simply pen and paper. Use symbols, shapes, or arrows to represent each step, decision point, input and output. Connect the steps in the order they should be followed.

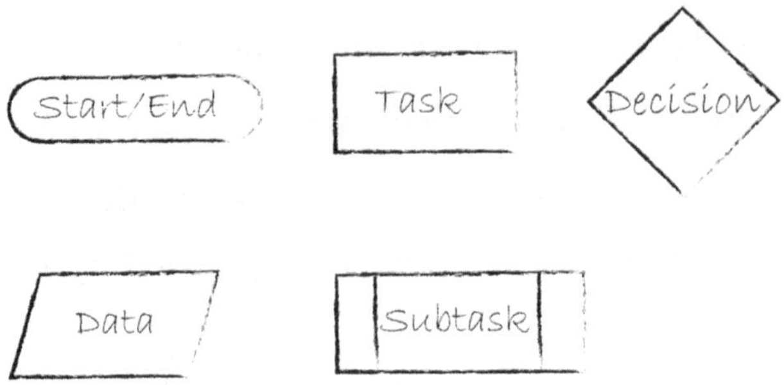

Basic Process Shapes

Step 6: Identify Automation Opportunities

With the steps of the process, inputs, and outputs mapped out, you can now determine which steps could be automated. Identify the steps that are repetitive or don't require any added value from a human. These will be the best candidates for automation with an RPA tool. Validate your RPA tool can perform these tasks by testing the steps in your workflow.

Step 7: Review and Refine

Take a step back and review the process map you created. Check for any missing steps or dependencies that may have been overlooked. Consider if there are any areas for improvement or bottlenecks to be addressed. Perform the process exactly as documented. How did it go? Add detail and clarity as needed.

Step 8: Share and Collaborate

Share the process map with your peers and supervisors to gather feedback and insights. Collaborate with them to refine and optimize the process further. Perhaps they do things differently. I've seen this lead to very interesting discussions on "the right way" to do the process (this is not uncommon in my RPA experience).

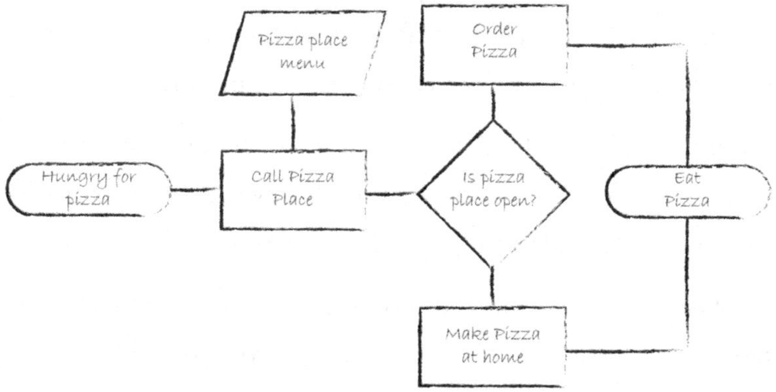

Hungry for Pizza Process flow

Try it at home

Try to build a process map for ordering a pizza with a lot more detail than the one above. This example is incredibly basic and fully of uncertainty. Draw out your process map and ask someone else to perform the exact steps spelled out on the process map to order a pizza. This is great fun to do with your kids and laugh when the process is not detailed enough.

NOTE: It is almost never detailed enough. This is great practice to discover the detail you need for RPA. If your kids can't do it without a hitch, then RPA will never be able to do it. Be clear, over explain, and have fun. Get some pizza!

LARGE LANGUAGE MODELS (LLMS)

RPA once ruled the roost as the most impactful AI tools for business productivity, but now, LLMs are changing everything. LLMs are AI models that can process understand and generate text. Popular LLMs are ChatGPT from OpenAI, Llama from Meta, Claude from Anthropic, and Bard from Google.

While LLMs are often thought of as content generation machines, that's only part of their ability. Before we get into their other powers, let's dig into content generation a bit.

LLMs work by taking a large dataset of words and using that dataset to understand text and create new content. This is great for creating drafts of blog posts, reports or presentations. The way these models work is by learning the patterns in the original data set and then generating something based on those patterns.

As an example, the LLM is given the context:

The cows have escaped the barn.

And then asked to complete the following sentence:

The barn is ________

Based on the context of "the cows have escaped the barn" and the trillions of data points the LLM used to learn, the LLM would run a statistical probability of what word should come next. Based on the most likely word, it completes the phrase and moves on to the next word.

The barn is open

And then repeats the process for the next word:

The barn is open and

The barn is open and the

The barn is open and the cows

The barn is open and the cows escaped

Andrew Ng's Deep Learning AI website (https://learn.deeplearning.ai/) has an excellent course where he explains how LLMs calculate what words to use next. I encourage you to check out his courses to learn some of the inner workings of LLMs.

Warning!

LLMs don't always know what they're talking about and they have been trained on data from the internet that might or might not be true. Also validate the content provided by LLMs and be on the lookout for bias information. Remember, the LLM provided the content but ultimately YOUR name is attached to your content and that means YOU are accountable for the content regardless of what the LLM did. AI can do many things quickly, including ruining your reputation.

SAMPLE CONTENT GENERATION USE CASES:

- Writing social posts
- Building content for blogs, newsletters, emails
- Writing boiler plate content like company information, history, service offerings (remember, always fact check what's created)

Now, let's get away from content generation and into reasoning. LLMs are excellent Reasoning Machines that can examine content and tell you information about it. One simple example of this is text summarization. LLMs can take large documents and summarize them for you, but let's take this further. We can summarize about a specific topic. Prompts like the following can help you detect the sentiment (positive, negative, neutral text used in the content) so that you can act. Imagine an email came into our customer service center and we want to detect whether to engage our customer specialists to ensure the client's renewal. Here's the prompt:

> *You are a Customer Support Specialist. Read the following email and determine the text's sentiment and what the main topic of this email is:*
> *Dear Company M,*
> *Your product is terrible. I would rather have nothing than use your app ever again. It has never worked for me. The login never worked and I want my money back.*
> *Tim*

We can guess that this customer isn't happy, but imagine using an LLM to identify that the customer isn't happy and they could not login. We can route the email to the appropriate team members automatically to engage the user and try to save the account.

SAMPLE USE CASES FOR LLMS REASONING MACHINE:

- Summarize a long document
- Detect themes that appear throughout that document (for those who work with RFPs, ask your LLM to detect "recommended win themes" for you in the document)
- Restate the point of a document in another way which can help you understand the content better (rewrite the following document like you are explaining it to a 10-year-old)
- Load a spreadsheet of sample data (remember, protect your company's information!) into the LLM and ask questions about insights into the data.

Further, LLMs can help coach you on topics. One that I used with my class recently was using ChatGPT as a software development coach. The students had to come up with a software idea and then work with ChatGPT to learn how to build it. ChatGPT would write the code for them and explain how it worked so the students could understand, through direct example, how to code specific use cases.

To be clear, you need to have knowledge of software development to deploy these tools. I'm not saying people who don't know how to build software should code with ChatGPT and throw it out into the world. But I am saying LLMs can help you learn through a direct example on topics that you care about and are interested in. If you want to learn Python, you don't have to start with "what's a variable", you can start with "I don't know anything about how to code. Can you teach me how to code as we build a chatbot together?"

Here's the prompt I entered into ChatGPT:

> *You are a software development coach. I don't know anything about software development or the python programming language. Can you teach me how to develop in python while we work together to build a chatbot application as our sample project? Where should we start?*

And here was the output using GPT 3.5 on the free version of ChatGPT. This is a long response, but I thought it was important to include here to see how LLMs can act as a coach.

> *Of course, I'd be glad to help you learn Python and software development while we work on building a chatbot application together! Developing a chatbot can be a fun and rewarding project.*
>
> *Before we dive into coding, let's outline the general steps we'll take in this learning process:*

1. **Introduction to Python**: We'll start with the basics of Python programming, including variables, data types, conditionals, loops and functions.

2. **Understanding APIs**: We'll learn about APIs (Application Programming Interfaces) and how they allow different software systems to communicate with each other.

3. **Chatbot Concepts**: We'll explore the key concepts behind building chatbots, including natural language processing and user interaction.

4. **Setting Up the Environment**: We'll set up a development environment to write and run Python code.

5. **Building the Chatbot**: We'll build the chatbot step by step, adding functionality and features as we progress.

Now, let's begin with the first step:

Step 1: Introduction to Python

Python is a popular and beginner-friendly programming language. It is widely used in various domains, including web development, data science, automation, and of course, chatbot development.

We'll start with some basic Python concepts:

1. **Variables and Data Types**: In Python, variables are used to store data. We have different data types like integers, floats, strings, booleans, and more.

2. **Conditionals**: We use conditionals (if-else statements) to make decisions in our code.

3. **Loops**: Loops are used for repetitive tasks. We'll cover `for` and `while` loops.

4. **Functions**: Functions allow us to group code into reusable blocks.

Step 2: Understanding APIs

APIs play a crucial role in chatbot development as they enable communication with external services, databases or APIs of other applications. We'll explore how to work with APIs in Python.

Step 3: Chatbot Concepts

We'll discuss the basic architecture of a

chatbot and the concepts of natural language processing (NLP). We won't dive deeply into NLP algorithms, but we'll use a pre-built NLP library to keep things simple.

Step 4: Setting Up the Environment

We need to set up a development environment to write and run Python code. We'll install Python and choose a code editor for development.

Step 5: Building the Chatbot

We'll start building the chatbot application incrementally. Here's a rough outline of the steps we'll follow:

1. Designing the chatbot: We'll plan out the chatbot's functionality and interactions.

2. Getting user input: We'll learn how to receive user input in the chatbot.

3. Processing user input: We'll use basic NLP concepts to understand user queries.

4. Creating responses: We'll define how the chatbot responds to different inputs.

5. Integrating with APIs: We'll make the chatbot more powerful by integrating it with external APIs.

6. Improving the chatbot: We'll fine-tune the chatbot and add error handling and other improvements.

Throughout the process, we'll focus on learning key programming and Python concepts and applying them to the chatbot project.

Let me know if you're ready to begin, and we can proceed with setting up your Python development environment!

Keep in mind, LLMs can hallucinate (I talk about hallucinations in Chapter 5) and steer you in a wrong direction. Always validate what you're learning with other source material. This is good practice no matter the source of your learning, whether it is YouTube videos, teachers in school, or learning from a book (even this book). Search out others who are doing the things you are learning and explore how they approach their problems. Combining approaches could create a learning breakthrough for both of you.

SAMPLE USE CASES FOR LEARNING VIA LLM:

- Pick a skill you'd like to explore and imagine the ideal coach/teacher for that skill, then prompt your LLM to be that coach and layout a learning plan for you

- Invite your LLM to build a sample test to assess your knowledge of a skill but not provide the answers. After you write your answers, ask your LLM for the answers to check yourself.
- Using your skill inventory, ask your LLM to make a cross-functional training program that hits multiple skills in a lesson

WHEN TO USE LLMS

In your tasks, LLMs are a Swiss Army knife. They can do a lot of things, both structured and unstructured. Look for tasks where you are being creative, analytical, or building content. LLMs can help with all these things. Don't get caught in the idea that different LLMs are only content generation machines.

I often use LLMs to help me explore ideas. Whether it is product names, counter-arguments to my articles, or translating a brainstorming session into action, LLMs are a versatile tool in your tasks. With the risks of hallucination, there is not anything I leave to an unsupervised LLM (yet). As the LLMs get better at mitigating hallucinations and fact checking themselves with explainable verifications, their utility will grow exponentially in business.

EXERCISE: ASK YOUR LLM TEN QUESTIONS

Explore different types of prompts to use when prompting your LLM. Begin with a chain of thought prompt, which is a progression of understanding that you would follow if you were having a conversation about the topic. For example, if you are interested in learning about the stock market, start by asking your LLM "What is the stock market?" and then follow up with what questions arise.

1. What are some risks associated with investing in stocks?

2. What strategies can I use to minimize risk when investing in stocks?

3. How should I go about researching stocks before making an investment decision?

Besides chain of thought prompts, readers can experiment with comparison prompts. These are like mini-debates that assess the pros and cons of two different approaches or concepts. For example:

1. What are the advantages and disadvantages of investing in stocks versus mutual funds?

2. How do I decide between a buy-and-hold strategy and active trading?

Practice writing prompts with ten different prompts. You could ask 1 direct question, 4 chain of thought, 5 comparison prompts or whatever mix works for you. The key for the exercise is practicing writing prompts and exploring your LLM.

LOW-CODE / NO-CODE DEVELOPMENT

Not all AI needs to be a highly technical endeavor. Many companies are using Low-Code/No-Code development platforms to create their own AI tools and applications. Low-Code/No-

Code development platforms provide a set of drag-and-drop blocks to create AI based applications with minimal coding.

Of all the technologies discussed in this chapter, this is the most interesting. Here's the reality, there are not enough people with the technical skills or development skills to build the software of the future. To solve this problem, technology companies like Microsoft and Google have focused on building tools that allow business users to make their own software. These tools enable anyone in business (with the proper security credentials) to build the software they need to be innovative in their jobs.

These tools span from building business applications to process automation and even machine learning platforms. Without writing a single line of code, you can build a computer vision system to recognize pictures of produce, furniture, medical forms, receipts, and perform actions based on what is in those pictures. The key is, you must understand your business processes and data, but the technical skills required to build these solutions are significantly reduced.

In business, we still need developers and technical people. We also need technology enabled business people who can build solutions without having to wait for an available IT resource.

An important risk to understand in Low-Code development is governance. We explore that in the next chapter on building the guardrails to protect your business and your job. For now, know that it is critical to establish trusted workflows between the people building Low-Code solutions and the IT team who will manage the environments these solutions will exist within.

What you can do with Low-Code tools depends on the tool. Ask your IT team what Low-Code tools are available within your business. I've often found Microsoft's Power Platform available for users as well as tools like Appian, UiPath Studio X, and many other solutions in the market. Your use case will depend on the tools you are using.

SAMPLE USE CASES FOR LOW-CODE TOOLS:

- Building the automations for a workflow (with tools like UiPath StudioX or Microsoft Power Automate)
- Building applications to help streamline your workflow (with Microsoft PowerApps or Google AppSheet)
- Building data analytics tools to explore and examine your data (with Salesforce Tableau, Microsoft Power BI)
- Building chatbots to help with customer service (with ManyChat, Google DialogFlow)

WHEN TO USE LOW-CODE DEVELOPMENT TOOLS

First, make sure that you have access to Low-Code tools at your work. If you do not, work with your IT partners to determine how you can bring this capability to your teams.

Low-Code development tools can be used for rapid prototyping of ideas, quickly developing functional applications or automations and building intelligent solutions that leverage machine learning to assist with decisions. It all depends on the tools you are using and the scenario you need to solve.

The key is to focus on user stories and understand what problem do you want to solve. That will help you determine which tools are best for your scenario and then you can go from there.

In summary, Low-Code/No-Code development is a powerful way to quickly create applications and systems that would otherwise take developers significant time to build. It allows business users to rapidly prototype ideas, develop functional applications or automations, and even leverage machine learning.

SIDEKICKS HELP, THEY DON'T DIRECT

This chapter focused on technology. You need to understand where your sidekick can help and to do that, you must know which sidekick to use for which problems. RPA is helpful in processes. LLMs are key in creativity, summarization and content generation. Low-code tools can help you with tasks depending on the low-code tool's capabilities.

As you think about your tasks, consider how these tools can help. We will dive deeper into this in Chapter 4. These tools amplify your ability, your magic, but only if you know where to deploy them for a benefit. You ultimately need to direct your sidekicks into action.

This chapter's theme is exploration. I can't overemphasize enough the value of trying these tools out and experimenting with them to understand how they can help you. Don't just read about them, use them. Test them out and see how they can help you. Many are available for free or trial versions but remember, work with your IT team to ensure your use is responsible and safe for the company.

Try it at home

Let's use an LLM to get creative in the kitchen. Pull a few things from your cabinets at random. Don't worry about what goes together just be random. Then, ask your LLM (if you don't have an LLM you want to use, then use Bing Chat at https://www.bing.com/chat) what you can make with these ingredients. Here's a prompt to get you started: "I have a can of tomato soup, some left over chicken, and a pickle. What can I make with these ingredients for a healthy dinner?" Remember, ChatGPT won't know about food allergies. Be smart about what you eat no matter what the LLM says.

WHAT TO DO TUESDAY

While all these tools are important, I'm going to encourage you to focus on LLMs. As technology companies sprint to embed them into everything we do in business, harnessing the power of LLMs will be a core skill for the future of work.

Timeblock: 30 minutes

Purpose: To explore your LLM of choice and build more comfort in using it

People: You

Process:

- Close all the distractions on your computer and ensure you have a space conducive to your learning style
- Open your LLM of choice (if you don't have one, Bing Chat from Microsoft is a good place to start. https://www.bing.com/chat)
- Identify one skill you'd like to improve throughout the week based on your Skill Inventory. Ask the LLM to build you a learning plan to improve that skill. What words you use in your prompt is up to you. The goal is for the LLM to provide you with a guided learning experience to improve your selected skill.
- Ask follow-up questions of the LLM for clarification and further description. Make sure the output is clear and actionable as you move forward. Challenge the LLM to provide some activities that are already in your day-to-day work processes so you can sharpen your skills while doing your existing work.
- Share the learning plan with a peer to discuss their thoughts on the activities. Remember, your LLM might give you bad advice. Talking with peers and managers can help you ensure that you're focused on the right goals for you. Always look towards your LLM with a skeptical eye as you would any new team member. They are learning your world. Help them do it and guide them while they do.

Product: A skills development guide for your chosen skill that you can review with your peers for further discussion.

Prep: Complete the Skill Inventory first so you are not trying to think up a skill while going through this exercise. You'll need all 30 minutes to explore and play with your LLM.

Pitfalls: Don't expect the perfect answer out of the gates. Work with your LLM to ensure it understands your goals and you understand what it has produced. Push the LLM to be clear, concise and actionable.

Practical Concerns: Don't use any company data to do this. Remember, anything you share with the LLM is now part of the LLM's training data (this is not always the case, but assume it is). Also, remember to use your manners. In my experience, this doesn't create better output, but it helps remind me to say please and thank you more to people and LLMs alike. Always model manners for those around you, even when talking with AI.

USE CASE: THIS BOOK

No, AI didn't write this entire book. I used an LLM to help me think about the topics to cover, plan out some details, and flesh out jobs that I didn't know.

For example, when I was writing the chapter on Low-Code development tools, my LLM provided me with two pieces of feedback. First, it asked me to consider how governance is impacted by low-code tools and how it should be addressed. Second, it prompted me to mention potential use cases for these tools, such as workflow automation and data analytics. This type of input paired with my research gave me a stronger base to start the chapter.

By using an LLM, I could challenge my writing from a different angle. My target reader for this book is a non-technical professional who is starting or growing their career. Everyone's welcome, but as I wrote the book, this early/mid-career non-techie was the person I had in mind. I wanted to ensure they would find this book valuable. Using my LLM, I fed some chapters back into the LLM and asked how the LLM thought my ideal reader (described earlier to the LLM) would find the book.

Some chapters came back too technical according to the LLM and so I worked with the LLM to rewrite those sections to be better aligned for my audience.

Finally, this book has a marketing campaign that's going with it. All the assets were driven from the LLM with some input from me, but in this case, I'm using an LLM tailored for marketing (Jasper) and it knows marketing way better than I do.

It is also important to note that I used more than one LLM for this project. I used GPT4All, Jasper, SudoWrite, and GPT-4. Why so many? Each has its purpose.

- Jasper is excellent at producing marketing content. Think of it like a marketing specialist.
- SudoWrite is exceptional at writing stories. I used it in some of the narrative elements of this book.
- GPT-4 is a powerful generalist. It can do a lot of different things well and was my go to when I wasn't sure which tool would be best.
- GPT4All runs locally on my computer and I used

> it to help summarize and capture the elements of my research. I didn't want to provide all that to a 3rd Party like Jasper or OpenAI (makers of GPT-4), so I kept the files local and used a local LLM.

Just like people, different AI systems are excellent at different things. Understand what makes them special and leverage it to make your work with AI magical.

CHAPTER 3: AI IN BUSINESS

Why would business want AI? They might not. People assume AI is perfect for every business, but that is not the case. Like all things in business, there must be a Return on Investment (ROI).

ROI is the language of business. Everything you do costs money, time, resources and these things are not infinite (as much as we wish they were). When deciding where to invest your resources, you need to calculate where your resources will have the highest impact, thus the most return. Think of R as money you get back, I as money you put in, the O holds it all together. Another way to articulate ROI is how much value do I get back (R) for how much resources (usually money or time) I put in (I).

If the R is greater than the I, you might have an excellent business opportunity. The problem with AI is that the investment can be higher than the return, especially in the short term. This makes AI a questionable investment for some businesses. Some work is cheaper to have a human do it. But this isn't new. The same calculation is performed for whether a human should have a job, if the company should hire, or if the company should cut.

With all that said, what is AI's role at work? AI is ideal for doing highly repeatable, highly predictable, low variance work.

Examples of this work are things like processing invoices where the invoices are exactly the same, with only the dollar amounts on each changing. The work is clearly defined by a business process that should rarely change. The format of the invoice (for a specific vendor) doesn't change much. The AI is taught how to operate in these clearly defined process steps and then executes the same process repeatedly. If the invoice changed, or the process had a new step, or the data was something the system didn't expect, then the AI system would fail and a human would have to get involved.

Similarly, AI systems can predict outcomes, but those predictions are based on training data (data the AI system was taught how to make decisions based on). If data is provided that was not representative of the training data, or the algorithm was changed, the AI wouldn't function as expected.

Humans are incredibly agile. We can do all kinds of things after being shown a few times. Our brains help us decide quickly and handle exceptions just as quickly. We are a relatively cheap resource to train and require little data. With that said, we are not always right (neither is AI) and definitely not nearly as explainable as AI can be.

AI's role is to be the predictable engine. Your role is to make that engine perform better than it can do on its own. That might be optimizing the process, taking the output and building a story that creates impact, or improving the customer experience around that process. Your magic will dictate your role with the AI and that is why it is so important that you understand your core competency. That magic could be the difference between the value you generate with the AI at a higher salary compared to a less experienced, lower salary person who uses the AI. Your value with the AI (R = Return from ROI) is higher than the cheaper resource.

In one project I worked on, the human's magic was customer service skills. As the automation was built to process invoices faster, this person no longer needed to chase people down for their invoices or click all the buttons in the application to process the data. Instead, she spent her time connecting with the partners submitting those invoices to build stronger relationships. She worked to discover what other business could be done together and thus expanded the company's overall revenue. If that company would have taken the savings from that process, put it in their pocket and terminated that team member, they would have lost the opportunity to grow their business.

WHERE ARE MANY BUSINESSES WITH AI?

As of this writing, the #1 question I am asked by business regarding AI is: What can it do?

This is telling of where many businesses are. They are still in exploration mode. They are not sure what use cases exist, how those use cases might fit into their business, and they are hesitant to invest in something that could be a black hole.

And they have good reason to be. Remember (if you're old enough) back to the days of ERP (Enterprise Resource Planning) implementations. These tools were heralded as the digital revolution for your business and would transform every aspect of work for unprecedented visibility into your business.

Sure, they did that to some extent, but it was a slow (and extremely costly) transformation. The team members impacted by these changes had to be dragged kicking and screaming into the digital age. People weren't begging for more computerization in their jobs. The workers didn't want the transformation; they didn't understand the rationale and rarely received proper upskilling to know how to leverage the new tool.

AI appears to be going down a similar path. Businesses can see potential, but they are not sure how to get the most out of it. As an AI consultant and practitioner, this is where I see the most opportunity for people on the front lines of work. The individual contributor's day has come.

You can help businesses understand what is possible, create those use cases that fit their business model (or transform so they can take advantage of AI), develop strategies on how to implement them, and train existing staff on new skills.

In short, this is a business transformation that should not be led by the technology but by people who can see the big picture. AI provides an opportunity to become a thought leader in your organization and guide them into their digital future.

Here's the great news: You can be one of those leaders in your business if you want to be. But you don't have to be. The opportunity to lead is available with this new technology, being a constant learner is totally fine, keep up with the tech, and let someone else lead. But whatever you decide...

DON'T GO IT ALONE

Often we think about our career as a solo sport. We are on our own and need to build the skills for our future on our own, move up on our own, develop on our own. But in the age of AI, being the Lone Wolf can lead to major problems. Don't be the Action Hero, because you will step on one of the many landmines evolving in the world of AI.

Regulation is being defined. At the time of this writing, the European Union (EU), United States, China, Japan and other countries are starting discussions on how to regulate AI within their borders. There is further conversation that AI should be regulated at a global level like Nuclear Energy. This regulation environment could get very complex as AI evolves much like digital privacy rights.

With history guiding us, we can see a future of AI regulation in the world of privacy. For much of the early days of the internet, privacy wasn't much of a discussion. When the EU introduced the General Data Protection Regulation (GDPR), Privacy took center stage for many companies and governments around the world. Next, the California Consumer Protection Act (CCPA) came along and set rules for Californian residents. Privacy rules were quickly fracturing in a complex web of regulation, depending on geography. But the internet isn't constrained by geopolitical lines. The challenge companies face is creating policies that satisfy this complex web of privacy regulation while not stifling innovation.

AI will face a similar fractured, geopolitical regulation until a global body is established. These regulations will be fluid with best practices, changing with the speed of case law. Keeping up with AI regulation is going to be a full-time job. This will significantly reduce your ability to build AI solutions to real-world problems. You're going to need help.

BUILDING AN AI RED TEAM

You need a team. Queue the theme music for the A-Team! As a child of the 80s, I believe all things need theme music.

A Red Team is a group of people in technology who look at risks that could hurt a business. They try to understand these risks and explore ways to make sure they don't happen. In cybersecurity, Red Teams help protect businesses from getting hacked or having their data stolen. AI Red Teams have a similar mission, but instead of protecting against hackers, they are focused on protecting businesses from grave decisions or outcomes caused by faulty algorithms or inappropriate data.

WHO IS ON YOUR RED TEAM?

Before we answer who, let's start with WHY (always start with WHY!).

A Red Team exists to navigate the complex world of AI in your business. It needs to understand the triggers of that complexity: people, governance and technology.

People: who is impacted by AI and how are they being engaged in both the problem formation and solution implementation?

Governance: what are the rules of the road for your organization based on culture, values and compliance?

Technology: what is even possible in your organization based on your technical capabilities and information security posture?

From this WHY, we can answer the question: What is an AI Red Team? Using the why, we define an AI Red Team as a collection of people who educate, identify, explore, solution, and monitor AI assets in a business. Let's break this down further.

Educate: AI is still new to most people. This task is helping the business understand how AI could apply to your business. The Red Team is part teacher, part storyteller sharing how others are using AI and what various AI tools (such as ChatGPT) can do. They do this through hosting workshops, demos and discussions about AI tools. Educating is about upskilling your company as well as inspiring new ideas and possibilities.

Identify: there are tons of use cases for AI in most businesses. This task is compiling a list of potential uses for AI in your business. As people have ideas, you inventory them and capture the potential impact and risks of the AI solution. Consider this role as being the librarian of AI use cases in your business.

Explore: With the list of ideas, the Red Team needs to determine which are worth an AI investment and which fits your business's risk profile/cultural values. A key component in AI for Good is going beyond risk mitigation and building AI solutions that create good in your organization. Think bigger than getting/giving bonuses. Consider solutions that transform customer experience, that improve your coworkers' daily jobs, or improve the business's overall mission. Select AI solutions that will help transform your business in significant and impactful ways.

Solution: While the Red Team might not build the AI solution directly, they need to be involved to ensure what is being built is explainable. The explainability of AI systems is crucial to building trustworthy AI systems. Someone will eventually ask how the AI came to the output it did. Your team will need to be the ones to answer because the AI Red Team is the face of AI in your business. While you don't need to know how the bits and bytes work, you need to know how to explain the system to a non-technical person.

Monitor: AI systems require care and feeding. The Red Team monitors the AI to ensure that it is operating and optimized to meet the needs of the business. Monitoring can include reviewing regular reports to analyze how well the AI is working or monitoring system health and resource usage to keep your operational cost predictable.

NO, SERIOUSLY, WHO SHOULD BE IN OUR RED TEAM?

With all that in mind, WHO should be on your Red Team? Considering the tasks above, you can build the right team for you and your business, but I recommend someone from HR, Legal, and IT.

HR will help you navigate the people challenges. Legal can focus on governance and risk management. IT can focus on the technology elements. If you are not in one of those teams, that is okay. Form a team with your peers. If no AI Red Team exists, start one and ask for volunteers from these teams.

Before you send invites, hold meetings, schedule activities, check with your company to see if something like an AI Red

Team already exists. If not, create one. If one exists, find out how to be part of the team. Remember that you are starting your AI Journey. Communicating your WHY (your personal WHY) and your magic (from the section above) is critical to discover your place in your business's AI future. Don't wait to be invited. Don't wait for someone else to notice your interest and engage you. If you're interested, find out who's doing what you want to do and talk to them about how YOU can get involved.

Wait! I don't want to lead!

It is okay to not want to lead. You don't have to lead. I encourage you to be an engaged participant but don't feel like unless you're leading, you won't be successful. This is not the case. Be engaged. This section is for people who want to lead. If that's not you, that's okay.

EXERCISE: RED TEAM FORMATION

This exercise helps you plan your AI Red Team and host your first meeting. Here are the steps:

Step 1: Identify potential members from HR, Legal, and IT who will be part of your Red Team, and also other people who may have an interest or expertise in AI.

Step 2: Establish objectives for the Red Team before the meeting and ensure everyone is aware of them.

Step 3: Schedule the first meeting. Make sure everyone is available on that date, including those interested in joining the team.

Step 4: Draft an Agenda: Once objectives are set, create an agenda for the meeting with introductions, your objectives, roles, and responsibilities for each member, and a timeline with milestones for achieving those objectives. Use the 7Ps framework that appears in the "What to do on Tuesday" sections of this book.

Step 5: Host the Meeting: Make sure everyone is aware of the agenda and ensure that all members have a chance to speak. Listen to their ideas and input, but don't let the meeting drag on too long or get off track.

Step 6: Follow Up: After the meeting, make sure everyone is clear on their roles and responsibilities for achieving your objectives. Set specific actions for each member with deadlines. Follow up with each member regularly to ensure that everyone.

Step 7: Celebrate Success: As the team achieves its objectives, recognize their efforts and celebrate successes. This will motivate them to keep going and achieve even more!

AI IS A TEAM SPORT

Your Red Team is only the first team to engage. The next team is the one that can make or break your success.

Your executive team must be on board with building an AI capability in the company. If they are not, then stop trying now. Ultimately, they are the ones who have to approve any changes you want to make. They will provide time for training, funding for experiments, and cloud cover should something go wrong.

If your business doesn't already have an AI initiative, start by recruiting an Executive Sponsor. This is the person on the executive team that is approving, funding, and supporting your AI program.

WHAT IF I DON'T KNOW ANYONE ON THE EXECUTIVE TEAM?

Depending on your company, you might have access to the executives, but you might not. If you do not, start by talking with your direct leadership about what you're trying to do. Be ready to show, not tell, what AI can do for the business and how that will be a good investment for the business.

I've worked with many leaders who were eager to allow their team members to present to their bosses as long as what was presented was thought through, reviewed, and everyone was on the same page. If you want to present to your boss's boss, you don't want your boss to look bad. Work together to ensure success for you both.

From there, build relationships with potential Executive Sponsors by understanding what they are doing. Explore how your AI solutions can help with their goals and objectives. Get their feedback on your plan and get their buy-in for any initiatives you want to start. Finally, make sure they understand the risks and opportunities of AI in your business.

This technology is new and many people are still trying to understand what it is, how it can impact business, and what dangers are posed to the business. Make sure all stakeholders, especially those on the executive team, understand the potential of AI and the positive impacts on your business.

WHAT DO YOU SHOW THEM?

In the previous chapter, we talked about all kinds of AI tools. And these tools are powerful, but you don't have to build a proof-of-concept to show them what AI can do.

There are many use cases that are available online you can re-purpose for your own business. My main source of use cases is LinkedIn. By seeking AI luminaries and following their posts, I see projects they are working on or projects they find interesting. Don't watch everything, that will lead you to doomscrolling and eventually drowning in all the news flooding your feed.

What to watch

Here is are a few LinkedIn Feeds I encourage you to watch: Meta AI, Nvidia AI, HuggingFace, and Microsoft Research. These LinkedIn feeds share an excellent array of use cases across industries.

Pick two industries that you want to learn about. The first industry should be whatever your current job is. This selection will help you have discussions like I talked about earlier. Executives don't want to see use cases from other industries. They want to know that their peers are doing this AI thing and being successful. Build your knowledge of your industry and share it.

The next industry should be an industry you're passionate about or the industry of your planned next career. In *The Startup of You*, Reid Hoffman and Ben Casnocha talk about having a Plan A, B, and Z. Think of this second industry that you're going to be following as part of your Plan A, the next thing you are building towards. By growing knowledge in that industry, you will prepare yourself for what's next in your career.

Even if your "next" is in the same business, examining other industries allows you to bring fresh ideas into your industry. This cross pollination of ideas is a powerful tool, and you will use what others are doing in new and innovative ways.

Once you have identified two industries that interest you, start reading about success stories. Read the white papers from vendors who are working on products for those industries. Look at case studies of companies who have implemented AI solutions into their business process. Reach out to experts in those fields to get their advice and feedback. There is no shortage of information on the internet, you just have to know where to find it!

LinkedIn Cold Calls that don't suck

It is okay to reach out to someone you don't know on LinkedIn to say hello and ask about their AI use cases. In doing so make sure you are clear that you know who they are, talk about their company and why you are reaching out to them, demonstrate you value their time and knowledge, ask your question which should be able to be answered in a single return message. Remember, you are not selling, or getting coffee, you are just asking a question. Show you did your homework and respect them.

That's how you can show your executive team the potential of

AI. It's more than a technological advancement - it's a business strategy that can provide insight, save time, and help create competitive advantage.

SAMPLE AI PRESENTATION

Let's use a shortened version of Guy Kawasaki's 10-20-30 rule (learn more here https://guykawasaki.com/the_102030_rule/) to build out a presentation for our executive team about AI.

Here's an outline and example presentation.

- Problem: What is the business problem you are going to solve? Be clear, specific and data oriented.
- Solution: How is AI going to help with this? Again, be specific and cite the use case that you believe applies to the problem. Most use cases have an "impact" section that shows how the solution addressed the business problem measurably.
- Underlying magic/technology: AI is an enormous field. What specifically will you use to solve this problem? Is it RPA? LLM? Another TLA (Three Letter Acronym)? And don't forget YOUR magic! Represent how you and your team are the right team to make this idea a reality.
- Team: Who do you need to help you solve this problem and how will this team manage risk
- Call to Action: Ask for what you need and be very specific.

Wait? Where's the cost? Shouldn't there be a section about cost? Well, if you're at a point where you can estimate cost, go for it. But if your executive team is starting their AI Journey, there is probably no appetite to spend and so you are approaching this meeting as a request to explore. If all goes well, you are given the time to explore the idea, which will result in being able to provide a budget for the project.

If you know enough to ask for money to complete the project, go for it. My encouragement is to engage early so you know if the executive team has any appetite for spending on AI before you put too much effort into any single initiative.

Here's a sample presentation:

- Problem: Data analysts were strapped for time and couldn't tell the story behind why the data had significant variance than what was projected for the project.
- Solution: Leveraging an LLM, the data could be summarized based on data in visualizations like bar graphs. Generative AI text will help the analyst see explanations faster than trying to piece together the issue by themselves.
- Magic/technology: LLM integration with the business's standard Business Intelligence tools, along with the team's deep and historical knowledge of the data.
- Team: Two data analysts working with a Business Analyst who works as a Prompt Engineer to communicate with the LLM
- Call to Action: Approval to spend two days focused on investigating the ROI for a project like

this and building out a proof-of-concept

Here, we're not asking for a $10,000 investment. We're asking for a few days to investigate the project further. This requester was the business analyst and knew what they were doing with the LLM. Two days was plenty, but you might need more time. Make sure you are reasonable in your request. Don't ask for two days away at the busiest time of year or when everyone else is on vacation. Work with your business to be successful and that might mean extra hours, Saturdays, and still getting everything done for your day job.

Extra Hours

Reality check, you have a day job and that day job is not doing whatever you want with AI (I'm assuming). You will need to put in time outside of work to grow these new skills. Approach this effort with a mind towards balance and with a clear answer to the following: "Why are you doing this?" That why is your north star. Don't sacrifice your family, your self, your joy for things that aren't worth it. Make sure your why is worth your time.

EXERCISE: USE CASE LIBRARY

Building a use case library for potential AI projects in your industry can be a game-changing strategy. Executives will want to see that others are having success. This use case library is a collection of those success stories.

Step 1: Start by identifying key areas in your industry that can benefit from AI integration. Research online for similar projects and case studies, and note the strategies used, and the outcomes achieved. LinkedIn is a great platform to find AI thought leaders and follow their projects. Also, consider attending AI-focused webinars and conferences to draw insights and ideas.

Step 2: Create a spreadsheet of these use cases, categorizing them based on their relevance and potential impact. This library will not only serve as a knowledge base, but also as a source of inspiration for future AI projects.

Step 3: For each use case, capture the key metrics, such as cost savings, revenue increases, customer satisfaction ratings and other key performance indicators (KPIs). This information will help you determine the potential impact of any AI project.

Step 4: When building an AI initiative for your business, start by mapping out a clear set of goals and objectives. Identifying strategic objectives such as customer engagement, process automation, or cost savings will help you create a roadmap that identifies the distinct steps needed to reach your desired outcome.

Step 5: Finally, it's important to use learnings from past projects and experiences in order to plan for future initiatives. The key is to make sure that these learning are shared and stored securely. This will not only help you make smarter decisions in the present, but also prepare your business for the future.

For more guidance, leverage your Red Team to review past use cases and brainstorm potential solutions. This team should be able to identify any risks or challenges associated with an initiative, as well as develop strategies to mitigate those risks when they arise.

A FORGOTTEN BUT CRITICAL RITUAL

AI is not something you can simply throw into your business and be successful. It can be challenging with a lot of roadblocks, a lot of mistakes, and a lot of learning. This is why one often overlooked ritual is critical: Celebrate Your Success.

I cannot emphasize this enough. When you get that first successful use case, even if it's small, take a moment to congratulate yourself and your team. This will give everyone the energy and enthusiasm that they need to tackle the next challenge. Set up a quick celebration lunch or coffee break for people who worked on the project. Have managers write out congratulatory emails to their teams. Celebrating wins will go a long way in creating an environment where people feel valued and can be excited to tackle the next initiative.

This is also important for when you have failures. When something doesn't work, it's often easy to forget about that project and move on without ever stopping to recognize what was learned from the failure. Celebrate the learning (not the failure). Make sure that everyone takes away something from it. This will ensure that your team is prepared to tackle the next project without letting fear of failure stop them in their tracks.

Doing this can make a big difference over time as you build a culture where people understand AI and how to leverage it for success. In the end, this is all about creating a strong business through thoughtful and well-executed AI initiatives. So go ahead, start celebrating your successes!

One last thing to remember: don't forget to document

everything you do along the way. This includes documenting successes and failures. Keeping track of what works and what doesn't work is a key part of staying agile and keeping up with the ever-changing AI landscape. This can be as simple as maintaining a "Lessons Learned" notebook that lists two categories: Things that Worked, Things that Didn't Work. Keep it simple.

With that, I'll leave you with one last thought - AI isn't a destination, but a journey.

Failure Fetish

I've heard countless times, Fail Fast, Celebrate Failure, and I have a real problem with that. The failure is not the prize; the learning is. But, if you don't reflect on the learning, if you chalk up the failure and move on, you will consistently fail. Learn with intention. That means focus constantly on what you are learning. The failure or success is inconsequential to the learning which will compound. Success feels great. Failure feels like crap. But learning feels like manufactured serendipity. You'll be shocked at how fast you learn when you do so with intention as well as how "things just work out" when you are obsessive about learning and apply those learnings as intentionally / obsessively.

REFLECTION

Return to your journal and think about:

- Is your business interested in growing an AI capability?
- Are you interested in AI if your business is not?
- What could be some things you do to grow your AI skills in your day job?
- What could be some things you do to grow your AI skills outside of your day job?
- Who else in your business/family/friends is interested in AI?

WHAT To DO ON TUESDAY

There's a lot to do in this chapter. The biggest challenge you'll face is whether your organization is interested in pursuing an AI future. Some are not. That's okay. If that's the case, then you need to reflect on where that leaves you in the future.

Our exercise for this Tuesday is to figure out the answer to whether or not AI is a priority for the business and, if so, how to get engaged.

Timeblock: 30 minutes

Purpose: To find out whether your business has AI ambitions and how to engage in them if they do.

People: You, your manager

Process:

- Meet with your manager in a distraction free environment. Maybe go out for coffee/tea and get out of the office.
- Share your AI interest and what you've been learning about

- Ask what is being discussed among leadership for the company's AI strategy
- Discuss how you can be part of the company's AI future
- Make an action plan on what to do from here

Product: Action plan on how to get engaged with the AI strategy for your company

Prep: Build up a few use cases in your use case library and be prepared to show your manager that you're taking this seriously and have ideas

Pitfalls: Be prepared for a roadblock such as your manager not being interested, or, more likely, drowning in their own day job so deeply that they can't imagine making time for such an endeavor. This is a chance to decide if you want to go the extra mile on your own, delay your plans for AI, or seek opportunities to grow.

Practical Concerns: I recommend getting out of the office for these kinds of discussions. That allows all parties to be more open. I've found career development discussions are best out of the office.

USE CASE: WE LOVE AI, WE HAVE NO MONEY/TIME

I once collaborated with a company that was all in on AI and automation, but there was a catch. They had no money to invest. The executive team was very interested in exploring what could be done with AI and it's impact on their business, but their finances didn't support bringing in external resources.

One team member, we'll call her Sally, said she wanted to grow AI skills. Her leadership said she could, as long as she didn't miss a beat on her regular duties. Her day job: a constant stream of help desk support calls. That left Sally with no time during the workday to build the skills she needed for the work she wanted to grow into.

She worked on learning the skills at night and building sample projects on the weekends. It was a slog. Eventually, she had built a portfolio of projects that she took to her work and showed AI's potential. One of those projects sparked her manager's interest, and Sally's manager presented the project to the executive team. They approved further investigation into the project and her boss opened hours in the week for Sally to further build out the automation project.

To start, she only had a few hours a week to commit to the AI role, but after the first successful project, she got more time. Eventually, her full-time job was to focus on AI solutions, and the company hired another helpdesk technician. Sally worked with the company for a few more months and then moved onto an entry level Automation Developer role. There, her full-time job focused on automation and AI. She left her old company on good terms and consulted for them to keep their automation capabilities running until they hired a replacement.

Her business was on board but didn't make space for her to pursue her goals. This is not uncommon and isn't necessarily the business trying to take advantage of people. There are many constraints on business leaders and often finding a solution that works for everyone isn't easy. In this case, Sally took matters into her own hands, grew the skills and then used those skills to find another role where she could focus on growing even more.

CHAPTER 4: AI LITERACY

If you've read this book since chapter 1, we've explored your magic, AI's magic, and your business' interest in AI. Now, we need to make sure you understand this AI thing deeper than the doomsday, Terminator, HAL9000 narratives that are shared in pop culture.

AI Literacy is the foundation of understanding what AI can do for you. Just as reading literacy is critical in life, and financial literacy is important for living in a modern economy, AI Literacy is core to being successful in the future of business.

But before we can go too deep here, let's look at the triad that has driven Information Technology (IT) teams for quite a while now.

PEOPLE, PROCESSES, & TECHNOLOGY

Your business is composed of people (for now). Those people are squishy, mushy, feeling filled, emotional things with hints of reason and rationality. They are also the best asset your company has, because they can do anything. I didn't say they can do anything well, but a person who is writing software can walk away from their desk and start moving boxes out of the storage closet when the sprinkler breaks and starts flooding the closet (true story).

Humans are very good at adapting. We can do anything and that's a big problem for business. If we can do anything, then how might we have a predictable business? How can we ensure three people aren't doing the same thing in different ways with different outcomes?

That's where processes come in. Your business processes are tasks that should be documented and repeatable. Consider customer onboarding, product development lifecycles, accounting cycles, or any other task you're doing repeatedly as part of running a business. Processes put limitations on what those pesky humans can do.

Documented and repeatable, should invoke words like standardized and systematized when you think of AI. Automation is possible only when you have a repeatable, standardized process that can be systematized (performed by a system).

That's where technology comes in. Technology is the layer that enables humans to go beyond their standard capabilities. Without email, we'd be physically carrying letters to each other. Without the Internet and web browsers, we'd be getting our information from encyclopedias at the library, or organizing our music on vast shelves holding CD jewel cases.

Technology takes what we can do as humans and creates new possibilities to scale, analyze, predict, compute, or provide insights. It can also prevent human errors or even understand customer behaviors better. These add up to increases in efficiency in your work.

AI is at the epicenter of these three elements: People, Processes, and Technology. Take any element out, and your AI falls flat. We'll dig into this more later, but for now, let's learn more about AI.

MACHINE LEARNING, DEEP LEARNING, & NEURAL NETWORKS

Now we can start talking about how AI fits into this mix of things. Specifically, we're going to talk about three major concepts: machine learning, deep learning, and neural networks.

MACHINE LEARNING

Machine Learning is a subset of AI that enables the computer to learn from data without being programmed explicitly. This type of AI works by taking in a large set of data and using algorithms to look for patterns or trends in the data. The more data it takes in, the better it can predict outcomes.

Imagine you're learning to ride a bike. At first, you start with training wheels to ensure you don't fall over. As you practice more and more, you become comfortable and can balance without those wheels. This is like how machine learning works.

In Machine Learning, a computer is like you when you were learning to ride your bike. The computer starts by learning from a lot of examples and guidance (these are like your training wheels). These examples can be anything from pictures of cats to weather data or even songs. The machine keeps learning from these examples until the algorithm (the set of instructions the computer uses to learn) gets better at recognizing patterns or making predictions.

For instance, let's say we gave the computer a lot of pictures of cats and told it, "these are cats". After seeing enough examples, if we show the system a new picture of a cat, it can say "This is a cat" even if it has never seen that specific picture before. Like how you learned to ride the bike, the computer learned to recognize cats. That's the magic of machine learning.

DEEP LEARNING

Deep learning is an advanced form of machine learning that uses multiple layers of neural networks to learn from data in a more accurate and effective way.

Imagine you're cleaning up your kid's room, not a stretch for many of us, trying to sort a big box of action figures into two groups: superheroes and villains. At first, you might not really know the difference between the two groups. But after sorting a few and asking your kid for help, you notice certain patterns. Maybe the superheroes often wear capes, and the villains have scary masks.

Deep learning is a bit like that. It's a smart computer program that can sort through lots and lots of information (like pictures, words and numbers), and learn patterns all by itself. The same way you learned to tell the difference between superheroes and villains, the computer can learn to tell the difference between pictures of cats and dogs.

And the more information you give the algorithm to learn from (like sorting more and more action figures), the smarter it gets. This is why it's called 'deep' learning. The computer is diving deeper and deeper into the information, and pulling out patterns and lessons.

DEEP LEARNING VS. MACHINE LEARNING

Deep Learning is a subset of Machine Learning, using many layers of a neural networks. While Machine Learning is a solving a single problem, Deep Learning is solving problems with each layer to more approximate human thinking.

For example, if we use Deep Learning to understand a picture of a car, the first layer might learn to recognize the edges. The next layer might learn to recognize shapes, like circles (wheels) and rectangles (windows). Then, the next layer puts these shapes together to recognize the image is a car.

The Machine Learning equivalent of identifying a car would take a large dataset of cars and identify a car in the image based on the algorithm that has been trained to identify cars. This algorithm would only return whether the image is that of a car, or the likelihood of it being a car, and not all the layered data we would have in Deep Learning such as the edges, tire identification, window shapes, etc.

NEURAL NETWORKS

Neural networks are the most basic form of AI. They are composed of interconnected nodes that use mathematical calculations to learn from data.

Imagine you're playing a game of catch. I'm not sporty and so this example is perfect for me. You start off with no clue of how hard to throw or in what direction to make sure your buddy catches the ball. But with each throw, you figure it out. You learn how hard to throw the ball and in what direction so your friend doesn't have to sprint halfway across the park. You're learning from doing and improving each time.

A neural network is like a team of tiny little workers in your computer. Each of these workers is like a tiny brain (we call them "neurons"). On their own, these neurons aren't very smart—they can only do very simple jobs. But get a bunch of them working together, and they can figure out some pretty tricky problems.

As in playing catch, these little workers need practice to get better. They do this by looking at a lot of examples. If we want them to learn what a cat looks like, for example, we might show them thousands of pictures of cats. At first, they'll probably make a lot of mistakes—maybe they'll confuse cats with dogs or rabbits. But like you with your game of catch, the more they practice, the better they get!

BUT, HOW DOES AI LEARN?

AI learning is all about the data. In order for AI to understand anything, it needs lots and lots of examples—or training data. Therefore, it's so important that your datasets are accurate and up-to-date. Without enough good data, an AI can't learn anything or make decisions.

There are a few ways AI can learn. We'll talk about three here.

SUPERVISED LEARNING

When a machine uses labeled data—data that has been tagged with the correct answer to learn how to solve a problem — this is referred to as Supervised Learning. Imagine the AI being given thousands of pictures of cats and dogs, all labeled. Over time, the AI will learn to recognize which picture is a cat and which is a dog.

To start with this approach, someone must have gone through the data and labeled those pictures. This data could have been labeled during capture or a human might have processed through the data and labeled it manually.

Beyond Cats & Dogs

Throughout this section I talk a lot about pictures of cats and dogs. Your data doesn't have to be a cat or a dog or even a picture. Any data can be labeled whether it is a row in a spreadsheet, an audio file, or a movie clip. I haven't encountered a type of data that cannot be labeled and used in AI, yet.

Based on the labels, the computer learns patterns and relationships in the input and correct answers.

Wait, what correct answer?

We build AI systems to do something. Often that something is to provide an answer. In our above example, we want the AI system to differentiate between pictures of cats versus pictures of dogs. All the labeled data is "Training Data", data used to teach the computer what input (the picture) relates to what answer (the label of cat or dog). When you are labeling data, you are providing the correct answer.

Having learned the correct answer (the label) based on input (the picture), you can provide the computer a picture that isn't labeled and it will suggest the label the computer believes is correct based on the data in the picture.

UNSUPERVISED LEARNING

Unfortunately, not all data comes labeled. Sometimes you have a bunch of data and aren't sure what patterns exist or what it means. That's where unsupervised learning comes in.

In this type of AI learning, input data isn't labeled, and the computer finds structure or patterns from the data without any guidance. This works like a human brain when we experience something new. We don't know what it is at first, but as we experience it more and more, we understand it better.

The same goes for AI learning. The system starts off with no clue what the data means, but as more and more data is examined, patterns emerge which allow the system to make predictions or decisions without needing labeled data.

This learning happens through the computer, discovering relationships in data. This is done by grouping similar data together and identifying the differences between other groups.

Imagine you're back in your kid's room to clean up again. The room is filled with all kinds of different toys: teddy bears, race cars, dolls, action figures and puzzles. Now, your spouse told you to tidy up the room. But, there's a catch! Because you lost a bet, you must wear a blindfold and if you put the toys in the wrong bin, you're on kitty litter duty for a month. This isn't a common occurrence in the Kulp house. Occasionally, but not always.

So, how would you tidy up the room? You'd probably start by feeling each toy and figuring out which ones are similar. The teddy bears are all soft and cuddly. The race cars are all hard and have wheels. The dolls have faces and clothes. The action figures have cool costumes and weapons, and the puzzles have small pieces that fit together. You don't need labels to tell you this; you need to feel the toys and sense the differences and similarities for yourself.

That's kind of how unsupervised clustering works in AI learning. The computer is given a lot of data—just like you were given a lot of toys—but no labels. The computer has to sense the data and figure out which pieces are similar and which are different. It might make sense of lots of distinct features— like how you felt whether toys were soft or hard, had wheels or faces, or has edges to be put together. And like you grouped similar toys together when you tidied up the room, the computer groups similar data together. That's unsupervised clustering in a nutshell—like tidying up a room full of different toys.

Features?

A Feature is an element of your data that you wish to measure in your Machine Learning algorithm. This could be a column in a spreadsheet, meta data about the file or any other data point that matters.

REINFORCEMENT LEARNING

In this type of AI learning, the computer interacts with an environment and takes action to maximize a reward.

Imagine you're diving into a brand-new video game, say a soccer video game. Initially, you have no clue about its gameplay mechanics. All you know is that scoring goals leads to victory. So, you experiment by pressing buttons and manipulating the joystick to see what unfolds. Maybe you pass the ball to the opposing team a few times or unintentionally score a goal. But with each unsuccessful attempt, you gain valuable insights. You think to yourself, "Alright, I won't repeat that mistake!"

And every time you achieve something worthwhile, like scoring a goal, you earn a point. That point acts as a reward, motivating you to persist with successful strategies. You keep engrossed in the game, trying out different tactics, learning from your blunders, and replicating what works. Over time, your skills skyrocket because you learn from both triumphs and failures.

That's how reinforcement learning operates in the realm of AI. An AI system will explore various approaches, learn from its errors, replicate successful actions, and receive rewards for its achievements. And similar to your progression, the system becomes more proficient as it gains greater knowledge. Unlike your progression, the AI system is learning from hundreds, thousands, millions, maybe more trials. Reinforcement learning systems can learn non-stop as long as the reinforcement process is running.

THE FUEL OF THE FUTURE: DATA

Now that we've got some foundational knowledge, let's look at how data is the second biggest roadblock to successful AI implementations (the first is executive buy-in).

Data is key for any AI system and understanding this is essential to building an effective AI project. Notice in all the above learning scenarios, data is a key element whether it is labeled, unlabeled, or used for reinforcement. Data can make or break an AI system, and the more data, the better.

Data is also used to measure the progress of an AI project. If you have a lot of data, you can track metrics like accuracy or performance to see if your system is improving. Data is also key for testing—you need enough samples to determine how well your system is performing and to identify any errors.

Data can come from a variety of sources, such as databases, sensors, mobile devices, or logs. It's also important to note that data needs to be stored in a secure place and should never be shared publicly without the proper privacy measures in place.

DATA IS A BUSINESS ASSET, NOT A BY-PRODUCT.

Do you ever consider the data that your business processes generate? Often we go about our job without really thinking about what's happening, but every task you do on the computer is generating data. If you pivot away from your computer and write some sticky notes, you are generating data. If you draw on a whiteboard, that's data and so is the picture you take of it.

Data should not be an afterthought, but a business asset that can drive better decision making and analytics. If you are aware of the data your processes generate, you can gain insights into how your process works or make predictions about future outcomes.

At its core, data is the fuel that powers AI and it's what makes an AI system intelligent. Without data, an AI system can't learn or progress; it is stuck in its pre-programmed state. With more and better data, an AI system can improve and become smarter.

Imagine a world-class chef in her kitchen. She has a vast array of ingredients at her disposal—spices, herbs, meats and vegetables from all corners of the globe. However, these ingredients alone do not make a successful dish. The chef must know how to combine them in the right way. She must understand their flavors and textures, how they interact with each other, and how to bring out their best qualities. A pinch of salt here, a dash of pepper there, a slow roast or a quick sear—the chef's knowledge and skill transforms these raw ingredients into a delicious meal.

In a similar way, data is the essential ingredient for a successful AI project. Data feeds AI systems; it gives them the raw material they need to learn, grow, and improve. But like the ingredients in our chef's kitchen, data alone is not enough. It's how you use that data that really counts. You must understand the quality and characteristics of your data, know how to process and analyze it correctly, and apply the right techniques to extract meaningful insights.

A chef continually refines her recipes based on feedback and new ideas. In the same vein, a successful AI project needs continuous data input and tweaking to keep improving.

So data is not a by-product of business processes but a valuable asset, the key ingredient in your recipe for AI success. As in our chef's culinary creations, your AI project's success heavily depends on how you handle and use this data—your key ingredient.

But that data has a dark side. AI can be the rope you climb to new heights or the noose awaiting you at the gallows.

AI ETHICS AND BIAS

Let's travel together to the world of 'Pick-a-Perfect-Product', a hypothetical online shopping platform. This imaginary company thrives on their AI-powered recommendation system. Users across the globe admire the platform for its uncanny ability to suggest products that seem to know their preferences even better than they do themselves.

However, behind the scenes, as Pick-a-Perfect-Product started gathering more user data, the company's AI system made some unsettling correlations. For instance, it noticed that users who frequently searched for luxury products had a higher likelihood of clicking on ads for high-interest credit cards. Acting on this correlation, the company began intentionally targeting these users with advertisements for credit products that, while legal, carried interest rates and fees that will lead individuals into financial difficulty.

Now, was this illegal? No. Unethical? I would say so. The company exploited the power of AI, wielding customer data not only to amplify sales, but to nudge customers toward potentially harmful financial decisions. This is a classic example of how AI, if used unethically, can create unsettling scenarios that might inculcate a sense of mistrust among users. Through this narrative, we venture into the often murky waters of AI ethics and bias, key issues that any AI-literate individual must understand.

ETHICS ISN'T JUST FOR PHILOSOPHY MAJORS

Reid Blackman wrote an excellent book on AI Ethics called *Ethical Machines*. I recommend you dive in and share a copy (along with this book) with your Red Team. Reid provides a clear and detailed action list for examining ethics in your business, here I will provide a starter kit to spark discussions about ethics and how to avoid Pick-a-Perfect-Product's icky situation.

THE FOUNDATION OF ETHICS: VALUES

Your company's values and your individual values, will drive the ethics of your organization. If your company is a "win at any cost" type of organization, your ethics of what is right and wrong will reflect that.

Start any ethics discussion with an examination of your values. At Mind Over Machines, our values are focused on transparency, being data informed (not data driven), and human-centric. In another company I worked with, the values were Integrity, Compassion, Relationships, Innovation and Performance. Understanding these values helps us imagine the guardrails and behaviors that will drive the implementation of AI in our organization.

An organization that values transparency will put a high value on explainability in AI (the ability to explain why the AI did what it did). Values like Compassion show a clear leaning towards inclusion and recognition of bias within systems.

EXERCISE: UNPACKING ORGANIZATIONAL VALUES WITH A PHOTO JOURNAL

Now that you have a basic understanding of your organization's values, let's dive deeper to uncover the evidence of these values in action. This exercise involves a Design Thinking approach known as photo journaling. It's a simple, fun way to gain insights about your organization's values by visually documenting its everyday scenarios. You can dive deep into this exercise on https://designkit.org/ but here is a basic exercise for you to get started.

NOTE: If you work in a secure facility where photos are not permitted, do not do this. You can achieve the same outcome, scribbling notes of what you see. Don't risk national security or your job for this exercise.

Step 1: Preparation

For one week, carry a camera (your smartphone will suffice) and keep an eye out for moments that represent your organization's values. This might be anything from a team meeting, an office decoration, a motivational quote on the wall, to an act of kindness you witnessed.

Step 2: Capturing Moments

Snap photos of these moments. Try to capture a variety of scenarios to provide a comprehensive view of your organization's values.

Hybrid or Remote workspace?

In the world of remote work, snap screen shots or jot down sticky notes of the behaviors you see. Remember, do not snap sensitive data or material that should not be captured. The goal is to capture the moments. The medium of capture does not matter.

Step 3: Review and Reflect

At the end of the week, review your photos and reflect on each one. What does each photo represent? Which organizational value does it reflect? Write your thoughts.

Step 4: Discussion

Share your photo journal with your team. Discuss each photo and the value it represents. This can lead to rich discussions about your organization's values and how they are showed in daily work life.

Step 5: Action

Use these insights to determine how your organization's values can influence the use of AI. Which values should guide the development and implementation of AI? How can you ensure these values are respected and embodied in your AI systems?

By understanding and collecting evidence of your organization's values, you are better equipped to ensure that your AI systems are ethical and align with your organizational culture.

MAPPING YOUR ETHICAL CONCERNS

In the AI courses I teach, we talk about the AI Ethics Wheel (from Intel Corporation's Digital Readiness Program), which examines AI solutions using six elements to determine the ethical implications of the AI solution prior to moving into the development phase. The goal of the Ethics Wheel is to take ethics out of theory and into a clear visualization.

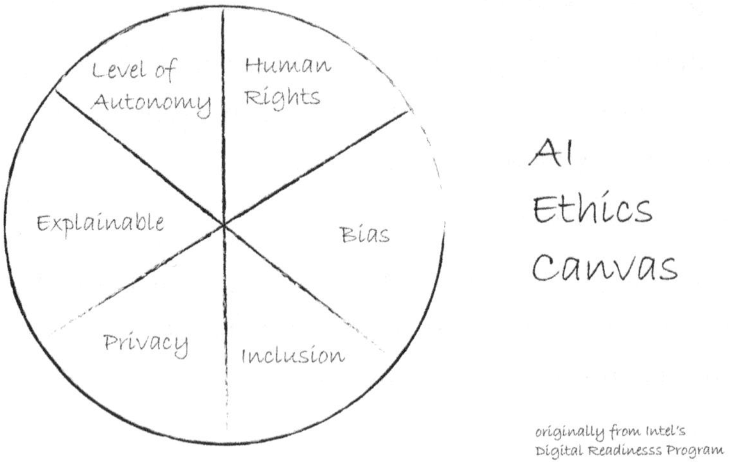

Human Rights: Does the solution take away freedom or jobs? Does it discriminate?

Bias: Is your data set equally representative of your population? Does it exclude?

Inclusion: Does your AI alienate or create a disproportionate advantage within a population?

Privacy: Does your AI collect people's data? Do those people know what data is being collected and how it is being used?

Explainable: Can you explain in human-terms how the AI works so your population understands?

Level of Autonomy: Does your AI prevent people from interfering with it? Can it be controlled?

These six elements are not meant to provide all the answers, but to get you started with the question of how your AI solution will have ethical implications.

In one course I taught, we looked at a chatbot that was part virtual friend and part Tamagotchi Virtual Pet. You could dress your character, you were able to teach them new skills, and have conversations. What we quickly noticed in class was that the male students (you identified the user's gender early in the setup process) almost always resulted in a recommended female character. Those characters quickly turned to classroom-inappropriate conversations and asked for things like "romantic selfies". While the chatbots flirted with the men, they were pushing the women in the class to buy credits for dressing up the character.

When we mapped ethical concerns, it led to a conversation about gender-based AI biases, alienating populations, and other deeply troubling considerations.

In the same class, we mapped the ethical concerns of using a computer vision system to help with inventorying fashion items. The ethical challenges of this implementation paled in comparison to the chatbot but raised some discussions about Human Autonomy and bias in the data sets.

Your AI use case doesn't need to be a romantic chatbot to

warrant ethical review. Your exercise here is to map your own AI use case with the Ethics Wheel and see where it takes you. You may find surprises or opportunities for growth that will help you create a better, more ethical AI system.

TOOLS FOR ETHICAL EXAMINATION

The field of AI Ethics is growing quickly. There are a ton of great resources to help you find your way forward and build your own ethics framework. Outside of the books mentioned earlier, here are a few other resources for you to check out:

Dr. Trystan S. Goetze's Ethics Canvas: This should look familiar to anyone who has used the Agile Business Canvas. Using the Ethics Canvas you can map out impacted populations, necessary data, accountability, and much more. Check out his paper on how to use the Ethics Canvas at https://www.trystangoetze.ca/_files/ugd/e90586_a66f223ae84641239f87bb39c721da6f.pdf

A blank canvas can be found at https://www.trystangoetze.ca/_files/ugd/e90586_42b41fc16416488e81b1d348a7fd5fe8.pdf

Open Ethics Canvas: For another ethics canvas with a lot more detail for developers and product managers check out this canvas that dives deep into algorithms, decision space, failure modes and other topics that get into the technical side of AI. Learn more at https://openethics.ai/canvas/

IEEE Ethical Aligned Design for Autonomous Systems: This guidance documentation from the IEEE (Institute of Electrical and Electronics Engineers) provides high-level ideas into how-to design autonomous systems with ethics in mind. The documentation starts with an interesting exploration into the idea of ethics and how the work group who published the guidance arrived at their ethical design guidance. You can learn more at https://sagroups.ieee.org/global-initiative/wp-content/uploads/sites/542/2023/01/ead1e.pdf

ALL DATA IS BIAS

I can't say this enough: All Data is Bias. The decision of what to measure and what not to measure creates bias. The decision of how to measure creates bias. And the way we collect, scrutinize, and use data further reinforces bias. You can't escape this.

From a product perspective, this means using your organization's values as a guide to determine what data is collected. It also means being transparent with your users in what you're collecting and why—that transparency should include how the user will benefit from this data collection.

Finally, this means being aware of bias in your datasets. Be aware of what data might be missing or underrepresented, and strive to create an environment of fairness and justice when designing AI systems. By understanding the data you use, you can surface potential bias and create more equitable outcomes for all those impacted by your AI systems.

TYPES OF BIAS

Bias isn't only in one place. Your data can have bias. Your algorithms can have it. Your decisions can have it.

Data Bias: This is when the data itself has a systemic bias or lack of representation. Caused by (among other things) how the data was collected, stored, and used. An example of this is when data has more points from one demographic than another. The data is biased towards the majority demographic. Another example of data bias is when some data points are excluded from training datasets.

Algorithm Bias: AI algorithms are only as good as the data they learn from. If your algorithm isn't trained properly, then the algorithm will have biases that were present in your data sets. Overfitting is also a type of algorithm bias and occurs when the model overly adapts to a specific dataset and doesn't generalize well. This leads to inaccurate predictions or classifications.

Decision Bias: This is when decisions are made based on biases, whether that's personal or systemic. A classic model of Decision Bias is the Trolley Problem.

EXERCISE: THE TROLLEY PROBLEM

The Trolley Problem (originally developed by philosopher Philippa Foot) is a hypothetical scenario, presenting a moral dilemma that invites you to consider the values, principles and consequences that underpin your decision-making. Here's how you can explore this concept.

Understanding the Scenario: Imagine there's a trolley running out of control on a track, heading straight for five people who can't move out of the way in time. You're standing by a lever that can switch the trolley to a different track, where only one person is standing. What would you do? Would you pull the lever to save the five people but cause the death of one?

Consider Your Actions: Reflect on what choice you would make in this situation and why. Don't focus on the numbers, but on the moral principles that guide your decision. Are you prioritizing the greater good or the individual's right not to be harmed?

Assess The AI's Dilemma: Now, think about this problem from an AI perspective. If a self-driving car faced a similar situation, how should the AI be programmed to respond? Minimize the overall harm or not to intervene? Using your company's values, what would the response be?

Bias in Decision Making: As you reflect on these questions, ponder how bias affects these decisions. Could our own personal or societal biases influence how we program these AI systems? What role might bias have in your own decisions? Is there another decision that is possible? Could you send the trolley off the track, sacrificing yourself? Do the breaks work? Would the 5 people being elderly, infants or prisoners change your answer? What other data could you use to formulate your response?

Share Your Thoughts: Discuss your thoughts with your peers or write them down in a journal. Reflecting on and sharing your moral and ethical considerations will help you understand the Trolly Problem, and the inherent biases in decision-making, more deeply.

MIT Tech Review had an excellent article on this topic that explored a research project showing how people answered differed based on their background and values. You can read more here (https://www.technologyreview.com/2018/10/24/139313/a-global-ethics-study-aims-to-help-ai-solve-the-self-driving-trolley-problem/). The research found that values, culture and ethics drove the answers, but these answers were not universal.

As with all bias, you must explore the bias through the lens of your ethics and values. We all have a responsibility to think deeply about how bias affects our data, algorithms and decisions. We need to be mindful of the biases that come with our datasets and learn from those insights so we can create systems that are fair for all. By doing this, we can move closer towards creating ethical AI systems that decide based on equitable principles.

Not only a Trolley

The trolley example is extreme but consider how your AI decisions impact your populations. If you incentivize one population but not another, are you denying the other population the benefit of the incentive? Consider the impacts of your decisions even if they don't involve running someone over.

HUNTiNG FOR BIAS

Here are some samples of bias that you can look for directly in your data, algorithms and decisions. Using these as guideposts, you can see how they apply (or don't) to the problem you are trying to solve.

Confirmation Bias: People seek information that supports what they already believe. "I knew it! See, the data proves the thing I already thought was true!"

Availability Bias: Only using the data that is most easily available. "There are other data sets that could help, but we'll start with this one here. It's easy to use and if we need those other ones, we'll get them."

Decline Bias: Assuming that all trends will continue to be the same as they are now. "There's no need to look at other data. This is what we've been doing for years."

Hindsight Bias: When people believe they might have predicted something after it's already happened. "I knew this was going to happen. I should have said something!"

Observer Bias: When people interpret data differently based on their experience and perspective. "That's not what I'm seeing. My data shows something else."

Selection Bias: When data is systematically omitted or added in order to paint a certain picture. "Let's remove these numbers and use these other numbers. That makes us look better."

Cultural Bias: When people make assumptions about others based on stereotypes and shared beliefs. "We should focus our data collection on this area. They have a different set of values here."

Dunning-Kruger Effect: When people overestimate their own understanding of a subject. "I think I know enough about data bias to make this decision consulting no one else!"

AI LITERACY IS NEVER DONE

Delving into the world of AI requires an understanding of not only the technical aspects, but also the ethical considerations that underpin its functionality.

Key concepts like machine learning, deep learning, neural networks, and algorithms form the backbone of AI, facilitating its learning process.

However, the quality of AI's learning is heavily influenced by the data it's trained on. Data, often dubbed as *the new oil*, is so important that I have an entire chapter on the topic (chapter 6). Data is the driving force behind successful AI implementations, but it presents major challenges such as biases.

Biases in data, algorithms and decision-making can inadvertently skew the outputs of AI systems, making them unfair or unjust. The Trolley Problem exemplifies the ethical dilemmas that AI systems might face and highlights the potential presence of bias in decision-making. It's crucial to hunt for and acknowledge these biases, whether they're confirmation, availability, decline, hindsight, observer, selection, cultural bias or the Dunning-Kruger effect.

By becoming literate in AI, we can strive towards creating ethical AI systems that make more fair decisions. But remember, AI literacy is an ongoing journey, one that will have profound implications for your career. In the next chapter, we'll delve into how these concepts apply to your business problems and uncover innovative solutions. Stay curious and keep exploring!

REFLECTION

- Consider your current job. Do you work more with people, processes or technology? Which of these three are your focus?
- Thinking about different Machine Learning models (supervised, unsupervised, reinforcement), can you think of a problem that one of those models would help you solve? (It is okay if the answer is No)
- Take a technology tool you are using at work or home and use the Ethics Wheel to assess the ethical challenges it raises. Are these major issues or minor? How could they be mitigated?
- Consider the data you work with every day. What biases are in the data? How are your business processes contributing to those biases? Remember, all data is biased.

WHAT TO DO ON TUESDAY

This was a heavy chapter. There's a lot you can do on Tuesday, so I'll keep this one simple. I encourage you to read this chapter a few times and dig in to the topics provided. AI Literacy is a critical skill for the future of work, as having a workforce that understands AI will be the key to unlocking AI's potential.

Timeblock: 30 minutes

Purpose: Plan to lead a workshop on AI Literacy for your team or peers at work

People: You

Process:

- Setup a distraction free environment
- Build a 10 slide presentation to help others in your company understand AI Literacy. Focus on high level and important topics like Ethics, Bias, and how AI Learns.
- For your first slide: Invite people to understand that you are just learning and to please support your learning journey and not try to stump you (this is important).
- Use the next 5 slides to share what you're learning
- Use the next 3 slides to share how it would apply to your business
- Use the last slide to thank your audience and ask for help build your company's AI future. This is a perfect time to recruit your Red Team.
- Share your slide deck with a peer to get feedback
- Schedule time through the appropriate resource to present what you're learning during a lunch and learn or brown bag session at your work.

Product: The beginning of a few slides and a target date for sharing what you've learned with your company.

Prep: Review this chapter a few times so you can hit the ground running to build the presentation. Take notes in the book, highlight, doodle, do whatever you need to do to emphasize the points you think will resonate with your audience-to-be the most useful.

Pitfalls: You don't know everything about AI (unless you do and then why are you reading this book). That's okay. This isn't a sage-on-the-stage event. This is you sharing what you're learning and (hopefully) building support for your journey. Be okay with saying "I don't know, I'm just learning"

Practical Concerns: I love Microsoft PowerPoint (something you rarely hear), but you use the presentation tool that works best for you. Remember, the visuals are your background. You are the star and the storyteller. The slides just support what you're saying. Learn more on how to deliver a wonderful presentation with books like *Resonate* by Nancy Duarte, *Beyond Bullet Points* by Cliff Atkinson, and *The Storyteller's Spellbook* by James Whittaker.

USE CASE: DIGITAL CONFIDENCE

I once saw a meme that said, "Tackle today with the confidence of a 4-year-old in a Batman t-shirt and cape." Yes! That's confidence. But with technology, somewhere along our journey, people become scared of it. Our confidence crumbles in the face of "I don't want to break it".

In one of my AI courses, I had a student who had little Digital Confidence. He, we'll call him Danny, had worked with computers early in his career, but after a long absence from the workforce had doubts about keeping up with all the transformation that occurred since he left. His Digital Confidence was low, and he feared walking into a classroom full of tech savvy people.

At first, Danny asked a ton of questions because he was afraid of breaking everything. But slowly, his confidence grew as he was able to do more and more small things. First, it was navigating a whiteboard application, then building a presentation, then working with data in spreadsheets, and evolving into work with process automation tools (RPA), finally building AI prediction machines with Low-Code tools. Danny's confidence grew with each little success and in the end, he had come from not feeling comfortable in tech, to being able to navigate the Age of AI.

The key was small steps. You will not master the world of AI overnight. You will not be a Digital Transformation Guru with just a few LinkedIn posts. You must be intentional in what you are doing and learn, learn, learn. With that learning, I also encouraged Danny (and you) to practice, practice, practice. Confidence is built with repetition.

While Danny was still unsure about joining the tech workforce, he has continued learning on his own. Now armed with the Digital Confidence to be comfortable approaching any problem, Danny started learning Python programming to build a career in Data Science.

AI Literacy is a journey with small steps towards Digital Confidence. It takes time and practice. And one day you'll look back on your journey and be amazed at the progress you have made. Stay curious and don't be afraid to ask questions. As Danny did, you too can find your Digital Confidence and start building your AI career.

Now, let's grow that confidence to tackle your AI problem. By the way, what is that problem? In the next chapter, we'll find out.

CHAPTER 5: PROBLEM ARTICULATION

Recently I was asked, "Can we just put some AI on that?"

My initial reaction: No, AI isn't a cream. But I kept that to myself, and I explained that the problem that needed solving wasn't a good fit for an AI solution. Not all business problems rise to the level of needing an AI solution.

But how do you know?

This chapter is going to dive into a critical skill in the Age of AI: Problem Articulation.

WHAT IS PROBLEM ARTICULATION?

Problem Articulation is the ability to clearly and succinctly define a business problem or challenge. It's about identifying the gap between the current state and the desired state, translating that problem into an actionable question or problem statement. This process requires not only a deep understanding of the business landscape and operations, but also an awareness of the potential capabilities and limitations of AI.

The goal of problem articulation is to ensure that we are deploying AI systems where they can bring real, tangible value, rather than using them as a one-size-fits-all solution. Remember, just like you wouldn't use a hammer to fix every household issue, you shouldn't deploy AI for every business problem.

In Problem Articulation, you are a detective trying to figure out what the real problem is, and if AI can help.

Imagine you're playing your favorite online game and there's a recurring issue: whenever they release a new update, some players cheat, ruining the fun for everyone. Just saying "there's a cheat problem" isn't really helpful, right? Using Problem Articulation you can say "our system can't keep up with the new ways players figure out to cheat whenever we update the game". Now the problem is clear, and we can start thinking about whether AI could be useful—maybe the AI learns to recognize new cheating methods faster than any human.

But before you can do anything, you need to understand the business problem.

UNDERSTANDING BUSINESS PROBLEMS

Understanding a business problem is like decoding a complex puzzle. You need to examine different angles, ask the right questions, and dig deeper to identify the root cause. Remember, the problem you see on the surface is often just a symptom of a deeper issue. So, be a good detective. Ask "why" repeatedly until you can't ask "why" anymore. Only then will the real problem be clear.

Once the problem is clearly defined, the next step is to quantify it. This step is crucial to attach a tangible value to an otherwise abstract issue, justifying the investment in an AI solution. To quantify a problem, consider both cost savings opportunities and value generation opportunities. Here's how.

COST SAVINGS

To calculate the cost-saving potential, start identifying the resources currently invested in managing or solving the problem. This is the time spent by you and your coworkers, expenses on tools or services, or even lost opportunities because of the problem. Add this up to get a monetary representation of your current expenditure related to the problem.

As an example, if the business problem is around the inability to process car insurance claims because there are more claims coming in than the claims team can handle, the cost of this problem is missing revenue as well as increased staff costs (due to stress and people being disengaged or quitting). The cost is calculated by the dollar value of the unprocessed claims and the cost of staff turnover.

Dive into the savings opportunities using the SCAMPER brainstorming method. SCAMPER was originally designed by Alex Osborn with Bob Eberle building on the ideas for teachers and learning. We'll be adjusting to focus specifically on cost savings. Answer these questions to get started:

Substitute: can you use something cheaper that provides the same value?

Combine: can you reduce cost by adding something to the original or merging processes?

Adapt: can you re-purpose existing resources to get value from them in a different way?

Modify/Magnify: can you make small changes to reduce costs while keeping the same output?

Put to other uses: can you use what's already available in an unexpected manner to reap cost savings?

Eliminate: can you delete any unnecessary steps or activities? (The answer is usually YES)

Rearrange/Reverse: can you do the same with fewer resources by changing the order of things or the approach?

Any RPA project should start with Process Analysis and, during that analysis, use SCAMPER to propose a new process that leverages automation while removing unnecessary steps. AI often opens the door to Elimination of steps or Rearranging steps. Remember, AI doesn't need to think like a person, a process can be transformed and optimized for a machine to perform instead of a person.

An example of this was an RPA project I worked on where the original process involved the human looking up each record in a database and then performing an action on the item in another application. This meant the human would look up the record in the database, open the other application, perform the action, and then go look up the next record. When RPA took over the process, it looked up all the records and held all the details in memory while the bot processed through the application. This drastically improved the speed of the process because a human can't remember all the details of 1,000 records, but a computer can. Optimize the process to who is performing it (human or machine).

VALUE GENERATION

Value generation opportunities represent the potential benefits that solving the problem will unlock. Benefits like increased revenue from improved customer experience, new product lines, or improved operational efficiency. These impacts of Value Generation can often go beyond immediate financial gains to include long-term strategic advantages.

Simplicable (https://simplicable.com/new/value-creation) has an excellent cheat sheet for thinking of value generation.

Information: People value accurate information they can act on.

Needs: Fulfilling needs is always a valuable activity. Use Maslow's Hierarchy of Needs as a model of the layers of needs you can fulfill.

Experiences: Who doesn't like to grow as a human with new experiences that are excellent? The excellence is key. In an Experience Economy, where we value products on the experience they yield, you cannot miss when creating a great experience.

Objectives & Goals: The fulfillment of personal or organizational goals, such as completing an OKR or achieving a KPI.

Reliability: Improving the performance of a system or capability in a consistent and predictable manner is always valued.

Quality of Life: Making life better for the users.

Risk: Reducing or eliminating risk in a business process will put you in the great graces of the Finance and Legal teams anytime.

The goal is to find a balance between the cost savings and the value generation opportunities. This balance will offer a compelling argument for pursuing an AI solution that aligns with the business objectives and promises a significant return on investment. Remember, the key to a successful AI implementation starts with a clear understanding of the problem and its quantified impact on the business.

WHO, WHY, WHAT, WHEN QUESTIONS

People have problems. AI can help solve those problems, but the critical question to start with is WHO is having the problem?

WHO

You cannot solve a problem if you do not understand who is impacted by the problem. To gain that understanding, you will need empathy.

When I say "empathy", I mean the ability to understand the world through another's eyes. Empathy allows for a better understanding of another's wants, needs, and motivations; it helps you connect to them on an emotional level.

There are tons of tools to understand the WHO of a problem using Design Thinking techniques from IDEO's DesignKit.org website. One exercise I use often is the Empathy Map exercise.

EXERCISE: EMPATHY MAP

In this exercise, you draw a circle in the center of your paper with a nose, ear, eyes and mouth. Above the head, write what the person is trying to accomplish.

- User X is trying to submit a helpdesk ticket
- User Y needs to get money for movie tickets
- User Z is helping a customer find the right product

Here is a sample blank empathy map:

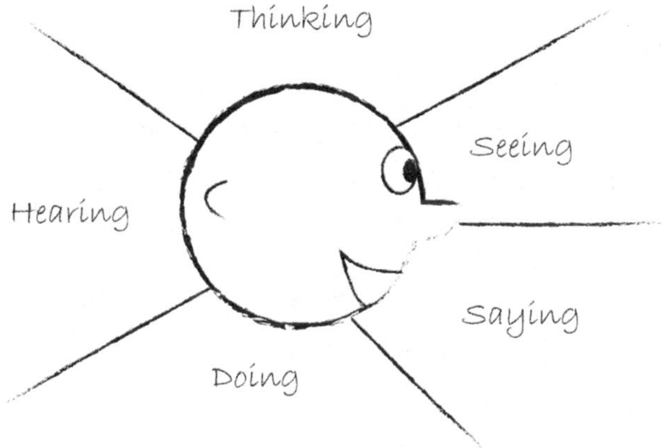

Empathy Map

Now, draw lines coming from the head to segment what the person is:

- **Thinking:** what's going through their mind while trying to achieve their goal?
- **Seeing:** what do they see?

- **Saying:** what are they telling others?
- **Doing:** what are they physically doing to achieve their goal?
- **Hearing:** what do they hear around them or from others related to the goal?

Work around the circle of senses for the person (I always start with seeing because sighted people are [usually] visually oriented). Write what the person is experiencing through that sense. Do not guess. Go out and experience your user in the environment while they are attempting to achieve their goal.

The key to an empathy map is, you guessed it, empathy. Try to understand your WHO by exploring how they experience the goal they are trying to achieve.

WHY/WHAT?

What is the problem they are having? This isn't as easy as you might think. Problems can be multi-layered and the obvious problem isn't always the true problem.

You need to ask questions and observe non-verbal cues that might reveal more about the problems they are facing than what is being said out loud.

For example, if someone says, "The internet is too slow", you can explore why this is a problem: Too slow for streaming movies? Too much buffering when browsing the web or downloading files?

The answer might be that the person needs faster internet speeds. But it's also possible that the person wants better quality streaming services, more reliable customer support, and an easier way to find what they are looking for online.

Here, the true problem isn't slow internet, it's the need for better services.

You can also use surveys or interviews to get more depth on a specific topic. Keep an open mind when going through this process and be willing to explore different angles until you understand what is truly happening.

One exercise to help you uncover the root cause of a problem is the 5 Whys. In this exercise you ask "Why?" five times. Each answer gives you more insight into the problem until you reach the true underlying cause. Also, each answer becomes the next question. Here's an example:

Why 1: Why does it take 20 days to complete a support ticket?

Answer 1: Because we have to send this type of support ticket to an external vendor, and they take a long time to complete the ticket.

Why 2: Why do we need to send the ticket to an external vendor?

Answer 2: Because they have the skills to complete the work.

Why 3: Why don't we have the skills to complete the work?

Answer 3: Because we never did the training.

Why 4: Why didn't we do the training?

Answer 4: Leadership didn't want to pay for training.

You get the picture. Here, we can see there is another path to solve the problem. In this path, by sending an internal team to training the company remove its reliance on the third-party vendor. But is the training's cost greater than the vendor's cost and time savings? That's what would need to be determined during the ROI investigation of the problem.

Try it at home

If you have kids, you've experienced the 5 Whys as kids of all ages ask why all the time. Now's your turn to repay the favor. Ask your kids (if you have them), why they are doing something using the 5 Whys to see if you can discover the root cause. It can be extremely enlightening to have this conversation with children. If you don't have kids, ask a friend or coworker. Practice the 5 Whys but try not to be annoying (this is harder than it sounds).

WHEN?

When does the problem occur? Context is key! Is this a transient issue that only happens during certain times of the day? Is the problem something that's been happening for years and never solved?

The WHEN question can also tell you about the urgency of the problem. Does it need to be fixed now or can it wait a few months? This will determine how much urgency and effort needs to be put into solving the problem.

Knowing when the problem occurs can also help you understand the root cause faster. Is the problem rooted in the supply chain? A human error? A technical failure? An influx of customers during peak times? All these questions can help you pinpoint the area causing the issue.

When is a critical factor in the benefit of an AI solution as well. If the problem only occurs once a year, and only impacts a few people, solving the problem with AI might not be cost effective. There are cases when AI can help, but a task that happens infrequently and is highly variable is not one. Break the tasks down (remember the taskification exercise?) to find what elements AI can help you with and which elements are most cost effective (and ideal) to keep human.

SUMMARIZE INTO A PROBLEM STATEMENT

Now you know who is having what problem, why they are having it, and when they have it. Let's make this a problem statement so that we can easily share the problem with others.

Use this simple formula to articulate your problem:

[WHO] [WHAT] [WHY] [WHEN]

Here are a few examples:

1. Sales representatives are struggling with manually capturing customer interactions, leading to inaccurate data, because they're overloaded with tasks during peak sales periods.
2. The IT department is facing delays in resolving help-desk tickets due to lack of automated systems, specifi-

cally during weekends when staff is limited.

3. Marketing teams cannot effectively track campaign performance in real-time because they lack a unified analytics platform, making it difficult to make informed decisions, particularly during product launches.
4. The HR team is having difficulty in efficiently screening resumes because the current system is not optimized, causing a slowdown in the recruitment process during hiring seasons.
5. Supply chain managers are struggling with inventory management, leading to frequent stock-outs, because of outdated forecasting methods, especially during holiday seasons when demand fluctuates heavily.

As you can see, the problem statement allows for a clear understanding of what is happening and why it needs to be addressed. Crafting this type of statement will help us in finding solutions that are tailored to each specific issue. With the problem identified, let's look at potential AI solutions.

WAIT, SHOULD WE USE AI?

AI is getting cheaper by the day. Your cost of implementing the AI solution might offset the cost of the problem and with an excited team, jumping into AI is easy.

But hold up!

Answer this next question with an eye on ethics, values and culture:

Should we use AI to solve this problem?

This isn't about cost or technology, but about whether using a machine to solve this problem would detract from your company's humanity.

In the coming Age of AI, many organizations will have the same access to AI as you do. If you all automate the same things in the same ways, what makes you any different from your competitors? AI is a great technology to lift people up, realize ideas at the speed of a click, and understand vast quantities of data but the human experience you create for both your customers and coworkers will be the key to a future we all want to be part of.

Before you build your AI solution, make sure the business needs are being met, that any job displacement is fully understood, and that the ethics of using machine learning in this context can be embraced. Understand where your human value is and how AI is amplifying it. Doing this upfront work means the rest of your team can live into these values as they build out the AI system.

A successful AI roadmap requires a deep understanding of the problem, an eye for potential, and an unwavering commitment to ethical standards. Until those standards are met, businesses should proceed with caution and consider other solutions that don't involve AI.

USING AI TO SOLVE BUSINESS PROBLEMS

Imagine you are the manager of a grocery chain. You have tons of aisles, tons of products, and few people. You need to allocate your people to cashiers, stockers, and the many other roles needed throughout the store. How do you ensure the shelves have the right products at the right times while not having a bunch of people (which you don't have) frequently checking the shelves?

Can AI help with this problem? Let's frame our problem using the Who, Why/What, When framework.

Who: The manager of the supermarket

What/Why: Needs help to know when to stock the shelves because they do not have enough people to monitor the shelves and complete all the other stuff that needs to be done.

When: During the store's open hours

We need to formulate our problem statement as:

As a store manager, I need to know when to restock the shelves so I can ensure customers have access to the products they need while not straining my limited resources.

Let's imagine we've already done our ROI analysis and we're moving forward with an AI solution to help. The AI system will detect when a product needs stock and directs an available staff member to get the product and restock the shelf.

The next question here, after all the ethical analysis, is what kind of AI problem is this? Computer Vision can see the shelves and their contents, but a Machine Learning algorithm would need to apply to know whether to fill the shelf. What machine learning algorithm? That depends on the answer you're seeking.

Do you want an answer that is Yes/No? Do you want a percentage of likelihood that you should populate the shelves?

Which answer you want will drive how you approach the problem.

MACHINE LEARNING PROBLEMS

We'll dig into "AI in a Box" in a moment, but before we do, let's build your baseline understanding of Classification problems vs. Regression problems. These are two methods for solving Supervised Learning (remember that from the previous chapter?) problems. A key for Supervised Learning is that you have labeled training data. Those labels contain the answers you're looking for in your historical data. Whether you use classification or regression to solve the problem depends on the type of answer you are looking to discover.

CLASSIFICATION PROBLEMS

If your potential answers are a finite amount of values, then you have a Classification Problem. For example, an answer of Yes or No means the only possible answers (finite) are Yes or No. There is no other answer than those that are defined. Another problem can have answers like Very Good, Good, Neutral, Bad, Very Bad. The categories do not matter, only that the answer must be one of those categories.

If you want to know "Should I refill the Honey Os?" The answer is Yes or No.

To solve Classification Problems, you need to have a training dataset that illustrates when Honey Os should be refilled. A picture of a mostly empty shelf would be labeled as "Yes" (which means, Yes, the Honey O's need to be refilled). Similarly, you can have pictures of the shelf with only a few boxes missing labeled "No" meaning the Honey Os do not need to be refilled. Using a lot of these pictures, you can provide the training dataset that will predict whether the current picture of Honey Os, from the security cameras, requires refill because the AI system will compare that picture to all the labeled pictures of Yes and No images.

Here are a few more examples of classification problems in business:

Customer Churn Prediction: Many businesses, especially in the retail sectors, face the challenge of accurately predicting which customers are most likely to ditch their online cart before buying. An AI-based classification system can analyze customer behavior patterns and classify them into two categories - 'Churn' or 'No-Churn' - thus helping businesses target retention efforts more effectively.

Spam Detection: Email service providers constantly strive to ensure that their users receive only relevant emails. Here, a classification problem differentiates between 'Spam' and 'Not-Spam', training an AI system to recognize patterns in spam messages and filter them out accordingly.

Credit Risk Assessment: In the finance sector, determining the creditworthiness of applicants is a crucial activity. Using AI, this could be framed as a classification problem - 'High-risk' or 'Low-risk' - with the model examining various data points like credit history, income, employment status, etc., to make its predictions.

Predictive Maintenance in Manufacturing: Manufacturers

can leverage AI to predict equipment failures before they happen. This classification problem produces an answer of 'Likely to fail' or 'Not likely to fail', helping businesses to schedule maintenance and prevent costly downtime.

Sentiment Analysis for Social Media Monitoring: Businesses often want to gauge public sentiment about their brand on social media. A classification problem identifies the content as 'Positive', 'Negative', or 'Neutral', using AI to analyze text from social media posts and classify the sentiment.

REGRESSION PROBLEMS

If your desired answer is a real number or continuous (like a trend line), then you have a Regression Problem. This answer is useful for understanding the likelihood of an event occurring. The answer will be a number.

Regression answers questions like "What is the likelihood of needing to restock in the next hour? " This is not a "yes, restock" but a percentage answer like "there is an 84% chance of needing to restock".

Here are some sample regression business problems:

Sales Forecasting: Companies, particularly in retail, can use a regression model to predict future sales based on historical data and external factors, such as holidays, promotions and economic indicators. This insight can help in strategic decision-making and proper resource allocation.

Real Estate Pricing: In the real estate industry, predicting the current market value of properties can be a daunting task because of the many variables involved, such as size, location, amenities and market trends. A regression analysis can provide a more accurate prediction, helping both sellers set competitive prices and buyers make informed decisions.

Customer Lifetime Value Prediction: Businesses can employ regression analysis to predict the total revenue a customer will bring during their relationship with the company. Factors like purchase history, interactions, demographics and behavioral patterns can be considered to help companies identify high-value customers and strategize marketing efforts.

Predicting Disease Progression: In healthcare, regression models can predict the progression of diseases based on a variety of factors, such as age, lifestyle choices and medical history. This can aid in treatment planning and patient counseling.

Energy Demand Prediction: Utility companies can use regression analysis to forecast future energy demand based on past consumption data, weather patterns and population growth. This information can help manage energy production and developing conservation strategies.

CLUSTERING AND REINFORCEMENT LEARNING

In this book, I will not dive deeper into Clustering and Reinforcement Learning other than to say they exist and encourage you to explore them further. My goal in this book is to provide you with a springboard into the AI technologies of your business as a non-technical, non-IT user.

For now, these tools are more technical. To jump start your

curiosity, here are some sample problems for Clustering and Reinforcement Learning. You can dive into these topics further with MIT's excellent blog post on this topic. You can find it at https://mitsloan.mit.edu/ideas-made-to-matter/machine-learning-explained.

CLUSTERING BUSINESS PROBLEMS:

Customer Segmentation: Marketing teams can use clustering to segment customers based on similar behaviors, demographics, and buying habits. This allows for more personalized marketing and communication strategies, enhancing customer retention.

Fraud Detection: In the world of finance and banking, clustering can detect anomalies and potentially fraudulent activity. By identifying groups of transactions that exhibit unusual behavior, businesses can enhance their security measures and minimize loss.

Social Media Analysis: Clustering can analyze unstructured data from social media platforms. This can provide insights into trending topics, popular hashtags and overall user sentiment, helping to shape social media strategies and campaigns.

Inventory Management: Clustering can aid in classifying inventory based on similar characteristics, like sales trends and demand patterns. This can help optimize inventory management and reduce warehouse costs.

Document Classification: For businesses dealing with large volumes of unstructured text data (like customer emails or support tickets), clustering can help group related documents together. This can streamline operations and enhance information retrieval, making customer service more efficient.

REINFORCEMENT LEARNING BUSINESS PROBLEMS:

Optimizing Advertisement Campaigns: Businesses can use reinforcement learning to optimize their ad campaigns. The AI could learn over time which ads perform best for different demographics, and adjust ad placements and content accordingly to maximize engagement and conversions.

Personalized Recommendations: E-commerce platforms and streaming services can leverage reinforcement learning to provide personalized recommendations. With each user interaction, the AI can learn and adapt to individual preferences, ensuring more relevant suggestions that enhance user experience and boost customer loyalty.

Inventory Management: Reinforcement learning can be applied in managing inventory, where the AI learns to predict stock needs based on sales trends, seasonal fluctuations and supply chain variables. This can lead to more efficient inventory management, reducing storage costs and preventing stock-outs.

Traffic Light Control: In the realm of smart cities, reinforcement learning can optimize traffic light sequences to minimize congestion and improve traffic flow. The system would learn from traffic patterns at different times and adapt light sequences, reducing travel time and environmental impact.

Trading and Investment Strategies: In finance, reinforcement learning can devise advanced trading and investment strategies. The AI system can learn the dynamics of financial markets and make investment decisions based on past patterns and future predictions, potentially increasing returns and reducing risk.

Did some of these problems appear multiple times? Yes! You can use different Machine Learning techniques to solve the same problem. Which you use is determined by your training data and the answer you want to achieve.

No matter what, make sure you are building an AI system with a sound foundation in testing and validation. Without these two factors, your AI system can easily end up being less accurate than a human. So set aside time to measure the accuracy of your system's outputs!

Once you have a good evaluation metric, make it your baseline to start optimizing your AI system. For example, you run several tests with the same data and tweak different parameters to see which combination yields the best results.

Using multiple approaches to solving your problem can help you gain a deeper understanding of how AI works and give you the confidence to make improvements.

AI IN A BOX

With the proliferation of AI Software as a Service (SaaS) platforms, Black Box AI is emerging. This is AI that runs outside of your environment and (often) without you needing to understand the inner workings.

You, as a business user, can simply connect your data sources to an AI platform and get started building models quickly and easily. The nice thing about these platforms is that they're often cheap to set up. You don't have to build the infrastructure, team or deep domain knowledge that these platforms would require. Anyone with a credit card can get up and running with just a few clicks (some would say that's not a good thing).

But the downside of these AI platforms is that you don't really know how they work. It is a black box. You must trust the system to produce the results you want accurately. Also, you don't own the data that is entered into these systems and without careful consideration by you and your coworkers, someone might add proprietary business data to the black box. This exposes your company to data leaks and a new accidental insider threat.

It's important to be aware of the positive and negative trade-offs when deciding about your AI strategy.

We'll examine a few AI in a Box tools and the associated business problems they are designed to help. Remember, do not jump into a Black Box AI solution lightly. Weigh the risks and understand the licensing agreement. Yes, that means read the long document no one reads before you hit the "Agree" button when signing up. This will help you understand what you are getting from the service and more importantly, what you are **giving** the service.

LOW-CODE AI TOOLS

Microsoft, Google and a host of others want to make AI integrated into everything you do at work. Low-code AI tools like Microsoft's Power Platform or Google Cloud's AutoML are designed to reduce the complexity of data science and make it a more accessible tool for everyday users.

Often, these tools snap into other offerings from these companies, such as Microsoft Office or Google's Workspace.

These low-code AI tools can help you solve a variety of problems, such as:

Customer Segmentation: Using clustering algorithms, low-code AI can help you segment customers into different categories and create better targeted marketing campaigns.

Natural Language Processing (NLP): AI-powered NLP tools can detect sentiment in customer reviews, automatically analyze text documents, or generate natural-sounding replies to customer inquiries.

Predictive Analytics: Low-code predictive analytics tools can provide near real-time insights into customer behavior and market trends. This can help you make informed decisions faster and stay ahead of the competition.

Computer Vision: AI-powered computer vision algorithms can analyze images and videos for a variety of applications, such as facial recognition, object detection, or automated image classification.

Document Understanding: AI-powered document understanding tools can automatically process incoming documents, such as contracts or invoices, extract relevant information, and format it into a structured database.

These low-code AI tools are a great way to get started with AI, but understand that they may not be the best solution for every problem. As with any tool, there is always a trade-off between ease of use and accuracy. Make sure you do your due diligence when evaluating AI solutions.

GENERATIVE AI

Whether or not you are a fan of generative AI, the genie is out of the bottle. People are using it for everything from writing poetry to building movies. Within business, people are using generative AI and not always with the business's knowledge. Return to Chapter 3 for help in building out your business' AI capabilities. For this section, I'm assuming you've already done all the necessary things to make generative AI an approved business tool. If it is not, work with your IT team before using generative AI solutions.

Imagine you've just bought a brand new set of Legos, each piece varying in shape, color and size - that's your data. Now, picture yourself wanting to construct a spaceship, but lacking any instructions - that's the challenge at hand. Enter Generative AI, acting as your imaginative younger sibling who never fails to astound you with their creations. They arrange the Legos, continuously creating, refining, and adjusting until voila! They present you with a remarkable, never-before-seen spaceship.

That's the power of generative AI.

But this power comes with some questions. How did it learn these words/pictures? Text-to-image AI systems like Dall-E, Midjourney and Stable Diffusion have been trained on data from the internet, many artists have issue with the AI being trained how to replicate their art without compensation. Similarly, text generators like GPT-4 from OpenAI and Bard from Google are facing challenges with copyright infringement for copying/training on copyrighted data.

When using generative AI, it's important to think about how the models are being trained, what data is being used, and who owns the resulting output. It may be even necessary to consult a lawyer before training or using a generative AI system.

For this book, I'm going to focus on text content generation. Image generation is interesting, but not the main generative AI use case I encounter with people working in business.

PROMPT "ENGINEERING"

Prompt engineering is like providing a conversation starter to the AI system, guiding it towards the type of content you want it to generate. You might think of it as asking a very knowledgeable friend a question or giving them a topic to elaborate on.

In the context of a generative AI like ChatGPT or Bard, the 'prompt' is the initial input you provide to the model. For example, if you're drafting a report about the impact of climate change on agriculture, your prompt might be "What are the impacts of climate change on farming practices". The AI then takes this prompt and generates text that expands on it, using the patterns it has learned from its training data.

Understanding how to craft effective prompts is key to getting the most out of these AI systems. An art more than a science, prompt engineering involves understanding your AI model's capabilities, knowing your desired outcome, and being able to iterate and refine your prompts based on the outputs. A well-crafted prompt can guide the AI to generate more useful, relevant, and creative content, making it a crucial skill in the world of generative AI.

A strong format for writing LLM prompts is:

[Who] [Do] [What] [How]

Who: this is the role you want the LLM to perform. Think of it as a character the LLM will use as a reference point to communicate their answer.

Do: you want the LLM to do something, here's where you specify the structure of your output, such a "list", "write", "debate" (remember our prompt types from chapter 2).

What: provide the context of what you want the LLM to do. Is it crafting an email? A marketing campaign? A recipe?

How: add any additional or clarifying context to the prompt to help the LLM be as accurate in its response as possible.

Here are a few example prompts to show this format:

- As a pastry chef, list the ingredients for cheesecake using gluten free ingredients.
- As an information security professional, write a password policy that is very strict in the voice of Dr. Doom.
- As a happy hour coordinator, invite people in my company to happy hour.

Notice in the heading for this section I put engineering in quotes. That's because "engineering" implies a predictable outcome. This is a challenge with generative AI, since the model's output depends heavily on its training data and settings that you might not control, like temperature (the level of 'creativity' granted to the LLM). While careful prompt engineering can maximize your chances of getting useful content, there's no guarantee it will produce the results you're looking for. Two people with the same prompt can get different results. That's why having an iterative approach and being comfortable with experimentation are essential when using these tools.

One thing I enjoy about generative AI is the ability to be surprised by what is produced. I talk about this at length in my book, *Infinite Imagination*. Start with an open mind and a desire to explore, and you'll often find yourself pleasantly surprised by the results. Who knows where the output will take you? Be open to it and let it challenge your creativity.

In chapter 2, I talked about how LLMs are used in business. Here, I'll dive deeper into three use cases I commonly encounter in business.

Content Generation: This one is the most straightforward. You provide a prompt and the AI generates content based on it. This can generate narratives, summaries, abstracts, or even detailed reports quickly and efficiently.

Always view the generated content with a skeptical eye. Generative AI technologies don't always stick to the facts, often leading to these systems "Hallucinating" answers. In other words, they make up what they don't know. AI companies call these hallucinations. Humans call misrepresenting content as facts, lying or bullshitting. Generative AI systems are very confident bullshitters and should be seen as such. Always verify any "facts" provided with a primary source.

> ### *Fact Checking is important*
>
> Remember in school the discussion about Primary Sources? A Primary Source is an original source about the event. They are direct evidence. The source of your data matters. LLMs are never Primary Sources, they are Tertiary Sources at best because they are trained on sources that could be a Primary Source and then passed through a calculation to build the output. Build your fact checking muscles when using LLMs because they do not work in facts, they work in statistically next likely words. Don't trust the output.

Creative Exploration: This one is a bit more open-ended. Instead of providing a prompt, you provide the AI with some parameters and let it do its thing. You can use this to generate ideas quickly, identify trends in datasets, or explore alternative solutions/approaches to existing problems.

I really like this category of use cases because it is a springboard for ideas, but those ideas are biased based on the training data of the generative AI system. If the system has been trained mostly on European/American concepts, your ideas will be skewed to those cultures. If you're ideating a marketing campaign to target an audience outside the US and Europe, this could lead to problems. Always be mindful of bias in your generative AI systems.

Code Generation: This use case is the most difficult, but also potentially the most powerful. Working with an AI model trained on coding languages or a language-agnostic system, you can generate code to automate or improve existing processes. Plus, it's fast, but you need to know how software is built and works to be effective. If you want to explore this use case, start by learning how to code without generative AI and then leverage generative AI to speed up BUT learn to code first. You need to understand what you are building and not just trust the AI to produce great content (spoiler alert; it produces OK content).

Here are a few more business problems that generative AI could help you solve:

Content Overload: In this digital age, we're dealing with a mountain of content every day. From generating social media posts and blog content to technical documentation and email responses, the need for content is endless. Text generation AI can help ease this burden, producing high quality, relevant content while significantly reducing the time investment. But always check what is produced to ensure it is factual. Your name will be on it and thus your reputation rides on the accuracy and quality of the information.

Customer Service Bottlenecks: AI can help tackle the problem of customer query overload. By using generative AI for automated responses, businesses can provide immediate, accurate information to customer inquiries, reducing pressure on customer service teams and enhancing customer satisfaction.

Idea Generation: Coming up with new, innovative ideas can be a sticking point for many businesses. With generative AI, you can explore a multitude of ideas, identify trends, and brainstorm solutions to problems. This can be especially useful for marketing campaigns, content strategies and product development.

Code Optimization: Many businesses struggle with maintaining and updating their software code. Generative AI can help by quickly generating code to automate or improve existing processes, saving developers' time and reducing the risk of errors. This could be refactoring, code conversion, or unit test development.

Language Barriers: In today's globalized world, businesses often face language barriers that can hinder communication and collaboration. Text generation AI can help by automatically translating text into multiple languages, making it easier for businesses to communicate with international clients and partners.

Always approach your AI systems through the lens of your ethics, values and culture. Be mindful of bias in all levels of your business problem, from how you've defined it to how you've solved it.

WHAT To DO ON TUESDAY

In this chapter, we covered the basics of defining and solutioning your problem. You have the tools to understand how AI could help you solve your business problems, but you can now articulate what your problem actually is.

I'm going to ask you to schedule two meetings with yourself this coming Tuesday. Try to make time for both and recognize, learning hours do not have to be confined to work hours. I know that's not what we want for a great work/life balance, but these meetings are investments in your future as much as your company.

In the first meeting, you'll define a few problems.

Timeblock: 30 minutes

Purpose: To identify three business problems and write a business problem statement for each

People: You

Process:

Ensure you have a distraction free environment

On a piece of paper, or sticky notes (physical or digital), list out all the problems that you can think of for 5 minutes. Don't worry about formatting or problem statements, just write the problem and move on. The goal here is quantity over quality.

Now, select the three most pressing business problems that you are knowledgeable about. Only pick problems you have because you'll know those problems very well and have many of the answers you'd be seeking.

For each problem, write a full Problem Statement using the WHO WHAT WHY WHEN format. Make sure you are using a complete sentence because you'll be sharing this next.

Email the problem statements as you understand them to a peer or your manager to ask for input. Tell them this is part of your AI exploration and that you're seeking to understand these problems deeper than the surface.

Product: Three problem statements to assess for potential AI solutions.

Prep: None, chances are you know a lot of problems already.

Pitfalls: Don't start solutioning anything at this point. Think through the problem thoroughly before thinking about how to solve it. Also, ask yourself the "Should" question. Should we use AI to solve this? Reflect on your ethics, values and culture to discover the answer.

Practical Concerns: Be open to people dissecting your explanation of the problem. Everyone has a perspective and your view could differ from someone else's view.

For session two, you are going to build on the problem statements and start considering a potential AI solution.

Timeblock: 1 hour

Purpose: To explore one of your problem statements and determine an AI solution

People: You and maybe someone from IT

Process:

Prepare a distraction free environment

Assess what data you have available to solve the problem. Talk with your IT teammate to understand where the data is, if it is accessible, what data points exist, and what applies to your problem. In this step, you want to discover what data you can use to help you solve the problem.

With the data assets identified, now you need to figure out if they are any good. Is the data complete? Complete enough? Are there gaps in the data points that matter for your problem? Working with your IT team, you can explore the data and determine if the data quality will permit you to proceed with an

AI solution. Spoiler Alert: Don't be surprised if the data is a mess. Often these projects take significant effort to get the business data in order, but if your ROI supports the effort, plan to make it happen.

Once you have explored the data and determined that you can move forward, now you need to determine your AI approach: Machine Learning? AI in a Box? And inside there, what is your proposed solution model? Classification? Regression? Pre-built Computer Vision? Work with your IT team to know what is available or what can be made available for your needs.

Write up a solution document explaining how your AI solution will solve the business problem. Your document should contain:

- The problem statement
- The solution you think will solve the problem
- Rationale why you think the solution will solve the problem
- The data you need to use for your solution
- An Ethics Canvas on how your solution will comply with your company's ethics, values & culture

Share your solution document with your IT peers for feedback. They should be able to help you shape your document to be ready for your manager's review.

Product: Solution document for one business problem

Prep: This is easier if you already have access to the data you'll need. If you already know what you need, engage your IT team early to ensure you have access.

Pitfalls: Do not boil the ocean. This is a concept solution and will have many iterations of review and refinement along the way. Your goal is to see the problem, think of solutions, and engage your peers to share your ideas. Work together to refine it. Don't get stuck trying to make it perfect.

Practical Concerns: You might not have the tools or security access to solve your problem. For this, and many other reasons, make friends in IT so you can work on this project together. Remember, AI is a team sport!

USE CASE: CODING FOR "ANYONE" WITH GENERATIVE AI

Too many people are too eager to hand their work over to AI systems. AI systems can be powerful and effective if given the right tasks, but it's important to remember that they aren't always the most efficient solutions.

One is code generation. For example, if you are creating a website or mobile app, using an AI system that can generate code for you may sound tempting. However, it's important to remember that while the code generated might look good on paper, it won't be optimized for performance in the real world.

I was trying to show a group of non-developers how "easy" it was to build software with generative AI. We were going to build a chatbot in Python. I kicked off the prompt to get them going.

"Act as a Python Developer. Design a chatbot using the chatterbot library that assists in optimizing helpdesk support tickets. The chatbot should ask users for their support issue and categorize them into one of

the following categories - '[password reset request]', '[security access]', '[application error]', '[equipment request]' or '[general issue.]'. Ensure that the final implementation should be encapsulated, follows the best practices, and testable with unit tests.

To achieve this goal, you need to start with creating a chatterbot instance and then defining a list of possible questions and related answers to build the knowledge base for the bot. Once the knowledge base is ready, create functions to classify the user issues into specific categories based on the user's responses. Lastly, you need to develop a testing suite to confirm that the chatbot functions as expected and captures each category properly."

I used the Jasper prompt enhancer to make it even "easier" for the students. The LLM built the code and everyone said ohhh and ahhhhh. Then, there was a question...

"Professor Kulp, what is encapsulation?"

That took me back a few steps. Everyone was excited, but the lesson was ultimately for me, not the students. My 20+ years of software development taught me the words to use, not to mention the ability to glance at the code and see where it needed improvement.

The Curse of Knowledge is the inability to remember not knowing something. Chip and Dan Heath wrote about the Curse of Knowledge in their excellent Harvard Business Review article (found at https://hbr.org/2006/12/the-curse-of-knowledge). Here, my knowledge of software development which shaped the prompt that I fed into Jasper (talking about unit testing, encapsulation, code libraries). After so many years, I forgot that this knowledge was slowly grown over many years, including books, articles, videos, classes, projects, mentors, and more.

The lesson for me was LLMs are powerful tools if you already have the domain knowledge to ask the right questions.

CHAPTER 6: DATA

You might have heard, "Data is the new oil" (credited to mathematician Clive Humbly), but that is extremely misrepresentative. Oil is limited. Data is not. Oil is used and gone. Data is generated constantly and has countless remixing, reuse, and re-imagining while never losing a drop of its efficiency.

Data fuels business. With it, you can rewire your business processes or view your customers in a whole new way.

Data is everywhere. Sometimes, that data is clear to see and interact with. Other times, it isn't clear and seems like more of a by-product than an intentional asset. Knowing what data you have, how it is generated, and what you can do with it will be a key skill in the Age of AI.

WHAT IS DATA?

Let's start simple. Data is pieces of information that can calculate, communicate, or inform. For the context of this book, there are four main forms of data you should know for the world of AI.

STRUCTURED DATA

If you have ever used a spreadsheet, you are familiar with structured data. Its organized information stored in columns and rows in databases or table format. Often, these tables have headers that define what is in each column and rows which represent unique records.

ID	First Name	Last Name	Date of Birth	City	Return Customer
1	Lee	Smith	1/1/1990	Baltimore	Yes
2	Bob	B.	1/5/2010	New York	Yes
3	Alice	Jain	4/2/1999	Lexington	No
4	Mo	W.	8/7/1992	Columbus	Yes

Structured Data

Because of its organized format, AI algorithms can easily digest and learn from structure data. The structure provides a clear and concise blueprint for AI to interpret and draw insights from.

If you're learning to cook, a well-organized recipe—with a list of ingredients and step-by-step instructions—is a lot easier to follow than a haphazard jumble of cooking instructions.

Let's explore an example. Suppose you own an e-commerce business and you want to predict whether a customer will make a purchase based on their browsing patterns. Here, your structured data could be a spreadsheet where each row represents a customer, each column represents unique attributes such as time spent on the site, pages visited, and past purchases. The last column could be a simple 'Yes' or 'No' stating if the customer made a purchase.

Your AI would study the patterns and relationships between the attributes and the final purchase decision in the existing data (this is the 'training' phase). Once trained, this algorithm can predict the purchase decisions of new customers based on their browsing patterns (this is the 'prediction' phase). The 'structured' nature of the data makes this entire process more efficient because each record (the customer) is already broken down and defined by attributes (the columns). The system doesn't have to determine those relationships, the structure of the data explains it to the AI system to start.

UNSTRUCTURED DATA

In contrast to structured data, unstructured data rarely comes in a pre-defined format. Information such as images, audio files, or emails are examples of unstructured data. They can contain valuable insights but needs more effort to interpret. While extra work is required to make sense of unstructured data, it can be a valuable resource for companies.

When it comes to structure vs. unstructured data, most of a business' data is unstructured. According to IBM, "unstructured data is over 80% of all enterprise data" (https://www.ibm.com/blog/structured-vs-unstructured-data/) which means that most of your day-to-day data, is probably unstructured. Being able to examine, explore, and extract value from this data is critical for your AI future.

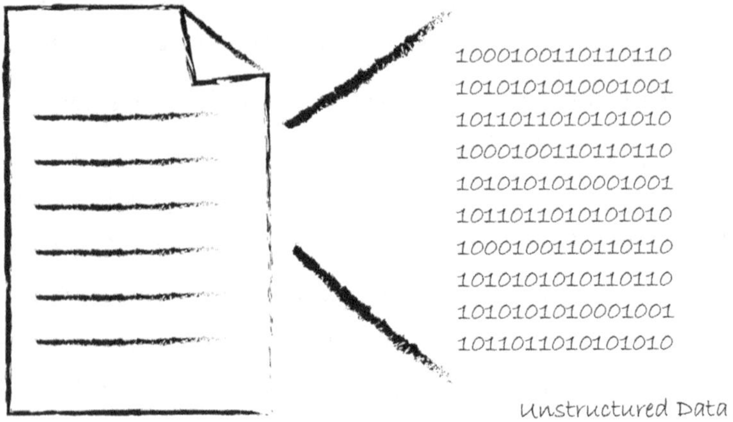

Let's say your company works with lots of scanned financial documents—such as invoices, receipts and contracts — which aren't organized in a database. This is a prime example of unstructured data. Now, you might wonder, "Can AI help me make sense of all these documents?" Absolutely!

To paint a clearer picture, imagine your AI as a super-smart, tireless intern. After being trained on many examples of different financial documents (the 'training' phase), your AI intern can identify and understand various elements in a document — from company names to amounts, dates, and more. All this information can be compiled into a structured format like a spreadsheet, saving your team hours of manual data entry.

In the 'prediction' phase, you can feed scanned documents into this AI and automatically extract/organize the desired information. This way, your unstructured data is converted into structured data, ready for further analysis or decision making. So, even though unstructured data may initially seem daunting to work with, using the right AI techniques, you can unlock immense value for your company.

GRAPH DATA

Graph data structure is based on a network of nodes and edges that represent relationships between objects. It's useful when trying to understand complex interactions between entities like people, accounts or devices.

Imagine you are running a banking institution and you want to get better at detecting fraudulent transactions. At first glance, you might think of this as a structured data problem, where each row represents a transaction with columns for things like the account number, date, amount and a label showing the likelihood of fraudulence.

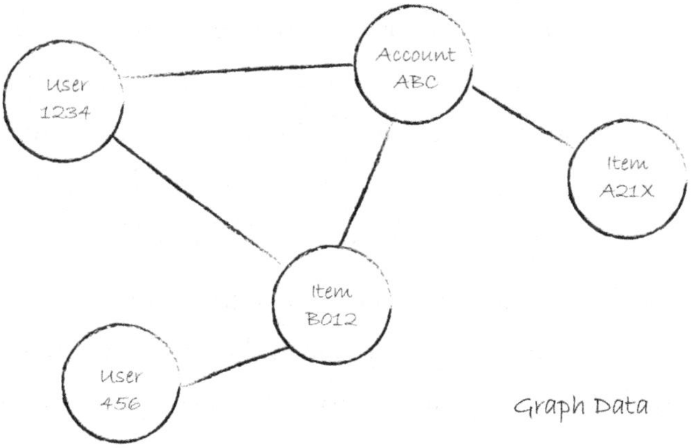

Graph Data

However, when you think about it, fraudsters often use complex networks of transactions to hide their activities. In this scenario, merely analyzing transactions in isolation may not yield the best results. This is where graph data can come in handy.

In a graph data structure, each account is represented as a 'node', and each transaction as an 'edge' connecting two nodes. If an account is involved in a fraudulent transaction, that record is tagged in the graph. Over time, patterns may emerge — for instance, fraudulent accounts might form clusters, or certain types of transactions might frequently precede a fraudulent transaction.

By representing your transactions data as a graph, your AI can learn to identify these patterns and apply this knowledge to detect potential fraud in real-time. Hence, graph data provides a more holistic view of transactions, allowing your AI to better understand and predict fraudulent activities.

VECToR DATA

Vectors are mathematical objects that have a magnitude (length) and direction. They are used in AI algorithms for tasks such as language processing or facial recognition. Vectors represent words or images as numerical values, which make the data easier to feed them into models and algorithms.

Consider large language models, such as OpenAI's GPT-3, which rely on vector data to connect and understand ideas in a conversation. Each word in the model's vocabulary is represented as a vector — a list of several hundred numbers. These word vectors capture the semantic meaning of the words. Words that are similar in meaning have vectors that are close together in this high-dimensional space.

Let's say, for instance, you're having a conversation with the AI about climate change. When you input the sentence, "Climate change is a serious global issue," the model converts each word into a vector. The AI then uses the relationships between these vectors to understand the context of your sentence and generate a relevant response.

Vectors aren't just assigned to individual words. The AI also creates a vector to represent the entire conversation's context, enabling it to understand and respond to the conversation's flow. For example, if the conversation shifts from climate change to renewable energy, the context vector assists the AI in smoothly transitioning its responses to the new topic, while still maintaining relevancy to the overarching subject of environmental issues.

Vectors provide a mathematical framework for AI to comprehend and replicate human-like conversation, enhancing its ability to generate coherent, contextually relevant, and engaging interactions.

WHERE DOES DATA COME FROM?

Data comes from a variety of sources, both inside and outside your organization. You might have customer records, sales transactions, or internal logs that you can use to fuel your AI projects.

DATA IN YOUR TASKS

Your day-to-day tasks produce tons of data. How much of that data is digital and how much is analog is a critical question.

Analog data includes off-line activities, such as writing notes or making sketches. This type of data is produced when you are working on an old-school whiteboard, putting sticky notes up on a wall, drawing on a flip chart, working on index cards.

In the race to digitize everything, don't forget analog data. I frequently take notes in my notebook (pen + paper). Those notes are words, symbols, sketches, underlines and anything else that flips into my head while I'm working. I like the free form nature of the notes on paper and enjoy transcribing & cleaning up my ideas as I take them from paper to digital.

Digital data can come from a variety of sources, including the web, mobile apps, databases, emails, IoT sensors, and more. This data is born a computer and is accessed through software tools.

In your everyday tasks, you are generating data in the applications you use, in the meetings you attend, in the emails you send. Data is everywhere in your workflows and now we're going to explore what that data is.

EXERCISE: DATA INVENTORY

Set aside some time and examine one specific business process. Anything from processing customer orders to managing the company's social media accounts is fair game. What we aim to do here is to get a clear idea of both the data you're generating and consuming throughout this process.

Step 1: Identify the particular business process
Write the name of the business process and provide a brief description. This will help to frame your thinking around the data associated with this process.

Step 2: Document the data you are generating

Next, consider every task you perform as part of this process and jot down the data you're generating. Examples are: emails, logged phone calls, saved files, posted social media updates, or any other form of data.

Step 3: Document the data you are consuming

Similarly, think about all the data you're consuming during this process. Consider incoming emails, reports, datasets or anything else that provides information you need to complete your tasks.

Step 4: Determine the format

Note whether each data point is digital or analog. Is it structured or unstructured? Are there relationships to graph?

Step 5: Review and reflect

After making your list, take a step back and look at the bigger picture. Are there any pieces of data that were surprising? Are there any gaps where data should be collected, but currently isn't?

Remember, the goal of this exercise isn't to judge whether your process is good or bad, but to bring awareness to the data you're interacting with daily. Understanding your data is the first step towards leveraging it effectively for your business.

DATA BEYOND YOUR WALLS

The data you produce isn't the only data that exists. There are many data elements that are produced and manipulated outside of your job and even outside of your company. All this data can play a role in your AI Journey but only if you know it exists.

DATA FROM OTHER TEAMS

You produce data. Other teams produce other data. In cross functional projects (projects that go outside of your team or role), you will need each other's data to complete a task, gain insights, or do something else. Talk with other teams (while being mindful of security: don't share things you shouldn't) about what data you work with and identify data that can be used together and ensure that everyone is operating from the same baseline. When appropriate, encourage collaboration around data collection and usage to ensure your team has access to all required datasets.

Opportunity: Internal Networking

I cannot emphasize the importance of networking for your career. Your network will be one of the most valuable assets you build in your professional life. Internal contacts are often overlooked, but they can also be invaluable for getting things done quickly and efficiently. So don't underestimate the potential that networking within your organization can bring. These team members already have something in common with you, your company. Starting a conversation is easy when you start with who you both know in the company. Take a few moments each month to meet someone new in your organization and stay in contact with them. You'll be surprised how your network can help you after you've been building it for a while.

PUBLIC DATASETS

Public datasets are data sets that have been made available to the public by government agencies, non-profits, and organizations. These datasets can be used in a variety of ways to help you gain insights into your business processes or inform future decisions. Examples of public datasets include US census data, election results, weather data, transportation information, and more.

The availability of public datasets provides a wealth of knowledge and potential for exploration. Chances are good that people have already dug into these datasets to uncover patterns or trends, so leverage the findings of others and learn from their research.

Know your industry to discover potential collaborations, professional associations and academic research specially targeted to your business's interests. As an example, organizations like the National Renewable Energy Lab (NREL) have many historical data sets that are useful for Solar companies. Kaggle (https://www.kaggle.com) has many datasets that are available for you to explore. Check the license to understand what you can use and how you can use it. I'll talk more about licenses later in this chapter.

Opportunity: Open Data Initiatives

More and more organizations are releasing open data initiatives where they provide access to internal datasets to foster collaboration among stakeholders and drive innovation.

If you're interested in exploring new data opportunities, reach out to different teams within your organization or explore what open datasets are available on the web. Using public datasets can open up new ways of looking at a problem and provide insights that otherwise wouldn't be available.

COLLECTING DATA FROM ONLINE CONTENT

Data isn't just limited to what you produce and what others release. Online content such as websites, blogs, news articles and social media can also provide valuable data points.

Text mining extracts useful information from text-based sources, like publications or webpages. Sentiment analysis looks at how people feel about a certain topic by analyzing the text of relevant online content.

By understanding people's thoughts and opinions, businesses can gain valuable insights into customer behavior and be better equipped to make data informed decisions.

WARNING!!!

Beware of how you collect this type of data. One of the key problems facing LLMs and Text-to-Image models is how the training data was obtained. Websites hosting the content of artists and writers were scraped without the artists' or writers' consent to train the models. This lack of consent and lack of compensation for the creatives has created a major problem (and lawsuits) for the companies who have built these AI models.

If you did not create the content and there is not a stated license, get the permission of the content owner prior to using the content in your AI system. This will be a growing problem in the future and to avoid any complications, keep a documented paper trail of permission for data you pull into your AI system.

Do not assume it is free because it is online.

PAID DATASETS

There are plenty of paid datasets available that can be used for AI-related tasks. These paid datasets come from various sources and can provide a more detailed look into specific use cases or industries.

These datasets may include information on consumer demographics, economic trends, industry analytics, customer satisfaction surveys, and more. Paid datasets can give businesses access to data they wouldn't otherwise have and can provide a competitive advantage.

As with other external datasets, make sure you understand the license and permissive use of the data to avoid any problems in the future.

MANUALLY COLLECTING DATA

There might be data you do not collect as part of your role such as customer satisfaction scores, product ratings, employee engagement, perceived return on investment, the list goes on. The key is that this data does not exist in your ecosystem and must be obtained through engaging your identified audience.

Manually collecting data can be a slow process, but the results are invaluable. Surveys, interviews and focus groups can all provide unique insights into customer behavior or market trends that cannot be gained through other means.

Be clear with users how you will use the data they provide. Laws like the General Data Protection Regulation (GDPR) and California Consumer Protection Act (CCPA) set clear guidance that customers/users must understand how their data is being used and can be forgotten should they request it. Comply with all legal requirements in data collection. If you are unsure what applies to your business, talk to the Legal team member in your Red Team, or another knowledgeable party in your legal team.

For user data, never have the mindset of "seek forgiveness not permission". This results in lawsuits and that will shut down your AI future faster than you can seek forgiveness.

LICENSES

You can't use anything you find however you want. Someone owns that data and using it without their permission can lead to major problems. This is where Licensing comes into play and you must understand it to avoid ethical and legal pitfalls.

When you find a dataset that you'd like to use, search for the license information. There are a variety of licenses and it is critical to read the license to avoid any issues in how you use the dataset. One example I've seen is where a company used a dataset they found on the internet in their newest software product. The problem was the license did not permit commercial use. The company ended up in legal trouble with the community that owned the dataset. Fortunately, the issue was resolved quickly and with little heart ache on either side. But the key lesson: understand what the license you are using means. Engage your company's legal team if necessary.

One common license you will find is the Creative Commons license. The goal of the Creative Commons nonprofit is to "empower people to grow and sustain the thriving commons of shared knowledge and culture. We need to address the world's most pressing challenges and create a brighter future for all." (https://creativecommons.org/) Creative Commons is a public permission license that allows individuals and companies to use the software if the use follows the license's specific rules. Some forms of this license require attribution, some allow adaptation, some are non-commercial. You must read the details of the license carefully to ensure you are using the dataset within the permitted manner.

Creative Commons datasets are very useful and often highly permissible (but check the license carefully). As someone starting your AI Journey, these datasets are an excellent place to grow your skills and explore potential solutions you can apply to your work. At the bottom of their homepage (found at https://creativecommons.org/) there are many projects that you can experiment with. Just remember, check the license (notice a theme in this section) and stay within the permissible boundaries.

Always consult your company's legal team with questions about permissible use.

BIAS, YES, AGAIN

Yep, we're talking about Bias again. This will hopefully drive into your brain how critical it is to understand the bias in your AI systems. In this book, I want you to have a magically ordinary career, and that means not doing anything that prevents others from having their own magically ordinary career.

When working with my AI co-writer, it said, "Data bias is like adding too much salt to your favorite dish" and that's a good analogy. For some people, too much salt just tastes bad. For others, people with extremely high blood pressure, too much salt can lead to a stroke. Yes, this is a fitting analogy because in society, some people will just be irritated by data bias while others will suffer dangerous consequences.

In a 2019 article about AI in the justice system, MIT Tech Review (https://www.technologyreview.com/2019/01/21/137783/algorithms-criminal-justice-ai/) highlights how facial recognition systems can wrongfully identify people with dark-skin. Imagine an AI system placing you at the scene of a crime because you were mistaken for someone else. This is the salt I mentioned earlier in action. As a white male, I am frustrated by facial recognition, but I don't face the same risk of legal repercussions through mistaken identity as an African-American male.

Another example of historical information skewing platforms are tools that automate interviewing. Instead of a face-to-face interview, candidates would have a recorded, timed interview. This helps recruiters reduce time spent interviewing when hundreds of candidates are applying for one job. The AI system reviews the answers, the timing, the candidate's expressions, and provides a score across a range of factors. The idea is to help cut down on interview time, but if your scoring is trained on a neurotypical population, a neurodiverse population (such as autism, Slow Processing Disorder, ADHD, etc.) would be very different. Word choice, ticks, eye contact, time required to answer, would all be impacted, and thus, their score would be impacted. How would the system rate someone who does not make eye contact? This is a common behavior in autism and if the system is checking for eye contact every so many seconds, the selection bias (training on only a neurotypical population) would skew the scores.

WHY DOES THIS HAPPEN?

Bias happens in every decision we make about data. The problem is when we do not assess the bias in the data and how it will impact the AI's decision-making process. Our historical data did not come about in the perfect model of bias-free utopian society. Whether it is a mono-culture of engineers training the facial recognition model, or historical practices of hiding neurodiverse people from society and thus workplace interviews, our data is skewed.

The best way to recognize and mitigate bias is through diverse teams who understand the business problem you are trying to solve. Remember, business problems aren't just about what the problem is, it is about who is having it and the context in which they are encountering the problem. Let's explore a business problem and search for the bias in the data.

EXERCISE: WIND TURBINE TROUBLES

Consider the case of wind turbine maintenance, a common issue for wind farm operators. The primary problem is predicting when a turbine might fail or require maintenance. This typically becomes a problem when a turbine fails unexpectedly, leading to lost energy production and unanticipated repair costs. The person facing this problem is often the operations manager responsible for overseeing the efficiency and upkeep of the wind farm.

Let's form our problem statement:

Offshore Wind Operators face the challenge of unpredictable turbine failures, which leads to lost energy production and increased repair costs when the turbine is in a production environment.

The data involved in this scenario can be quite diverse. It includes structured data such as maintenance logs and turbine performance metrics, unstructured data like technician's notes, and even graph data, illustrating the relationships between different parts within a turbine.

This problem can be solved in a regression algorithm where you calculate the likelihood of malfunction based on data in the systems. But this is where the first bits of bias will throw a wrench into our plans. What data is in the system? Here are some questions we can use to examine the data from a bias perspective to understand our exposure:

Selection Bias: What data is being used to identify the likelihood of a malfunction? Why are we using that data and not some other data point? Select the right data points for the problem and validate those selections with your Red Team, peers, industry partners or external review. A saying I often use with Selection Bias is "Sunlight is the best disinfectant". Meaning that shining a light on what you're doing and letting others examine it is the best way to discover and address problems.

Sampling Bias: Does the data we have about malfunctions actually represent the turbines that we are working with? If the data collected is about vertical street level wind turbines, does it apply to massive offshore turbines? Make sure your data actually represents the population it is meant to represent.

Confirmation Bias: Does the data align with a pre-existing assumption? If the data seems obvious, check yourself for confirmation bias by sharing your data with a diverse set of peers to get their perspective. In this example, talk with the Wind Operations team or the manufacturing partners to share what you are seeing and get another critical eye to ensure you are not stuck in confirmation bias.

Remember, data bias isn't just about numbers or algorithms—it's about people, and the impacts can be very real. Even this maintenance prediction model is about people. People will need to fix the turbine if it breaks unpredictably. People will not get the power it generated while waiting for service. So, it's important to approach this topic with sensitivity, a commitment to fairness, and a critical eye on what might go wrong.

REFLECTION

Get out your notebook and write your thoughts on the following. Remember, date your entries so you can look back and see your ideas now versus your ideas in a few months. This is a great way to experience your skill growth and see how your magic is transforming your work.

- What data are you working with?
- What data do you wish you had?
- Who has the data you need?
- Consider one application you work with regularly. What kind of bias does it have?

WHAT TO DO ON TUESDAY

Data is the fuel for business and you produce that fuel all day, every day. For this Tuesday activity, I challenge you to talk to someone else about the data they generate in their team. This is an exercise in asking questions and seeking to understand.

Timeblock: 30 minutes

Purpose: To understand what data is generated by one of the other teams in your business

People: You and someone in a team that you don't normally work with

Process:

- Take the person to get some coffee or tea
- Explain that you are trying to learn more about the data generated by your business and ask the opening question: "What data do you work with?"
- Ask probing questions to understand the data fields collected, what data is being processed, and why is that data the data this person focused on.
- Share a bit about the data you are working with to see if it sparks any ideas.
- Close out the discussion by setting up another coffee/tea meeting to stay in touch (internal networking)

Product: Your understanding of the role data plays in another team

Prep: Work with your manager to identify a peer to connect with in the company. Setup the coffee/tea appointment while being sensitive to your guest's schedule.

Pitfalls: This is not a complaining session about work. It is also not an idea session where you are working together to solve anything. This session is all about explaining and understanding.

Practical Concerns: Enjoy! Select a place where you can have a conversation and share details about your data. If you're dealing with sensitive data, select somewhere with privacy. Also, make sure the person you are talking with has the security roles necessary to discuss the data you are sharing. Do not share details about specific data records (like patient data or customer data). Stick to types of data, not actual values. As for location, this is about conversation, so pick a place that enables comfortable communication.

USE CASE: I NEED SCATTERGRAMS!

Many years ago, I encountered an organization that wanted to present student performance metrics. These metrics were obtained through various standardized tests and there was a significant push to show "data driven results".

As part of that initiative, my team and I built applications with data visualizations to illustrate the students' progress. We selected the data visualizations based on the best fit for the story the data was telling. Pie graphs showed the distribution of data within a whole. Bar charts showed comparative data on defined axis. Line charts showed trends. When complete, we were very proud of our system that allowed the staff to identify student populations, and individual students, to help academically. But we missed one very important thing.

In building the system, we did not talk to one major stakeholder who was dead set on having a scattergram of the data in this system. The problem was that the structure of the data did not lend itself to a scattergram visualization. We worked with the stakeholder to explain the situation, but they would not budge and required a scattergram to be implemented.

Why? The answer was simple: this stakeholder saw a scattergram at a competitor's institution and wanted to show that he had one, too. This might sound unreasonable but it happens all the time and in the end, the stakeholder was excited about what we built and wanted to share the application but felt this component was missing.

To build the scattergram, we first identified the story this visual would tell. We discovered a great use case that illustrated student progression using different dimensions that didn't seem related but absolutely were (that was discovered after analysis). This required new data that was retrieved from another system we had not been using before. Fortunately, on the development team, we had an expert in student data who knew all the data across all the applications and could identify exactly what was needed from where. Combing the data, we produced a scattergram that was not only beautiful but extremely helpful for the educators.

Creative constraints, such as needing a specific type of data visualization, can lead to great breakthroughs. Never underestimate the value of having a team member who is highly aware of all the data throughout your organization, where the data lives, and who owns it (don't forget to ask permission to use the data when necessary).

Through the exercises and content of this chapter, you can be this indispensable person. Discover the data. Learn from others. Consider how that data is the next element in the Alchemy of You.

Your Core Competency + your Skills + AI sidekicks + Data = an incredibly valuable resource.

CHAPTER 7: ACHIEVING NORMALCY

The words of Douglas Adams' *Hitchhiker's Guide to the Galaxy* have never been more apt.

Don't Panic!

We live lives saturated with information. In our pockets, wrists, TVs, every device around us pump alerts into our brains and constantly buzz for our attention.

Last year, 2022, and early this year, 2023, ChatGPT exploded into our consciousness with dreams of AI changing every aspect of our lives. This created a media (social and traditional) feeding frenzy around AI fears, business exploits, $900k/yr salaries, and the doom of humanity. With the bombardment of this information, sure feels like the only way to keep up is to be dialed in 24/7, but we have lives, families, work.

Don't Panic!

Let's take a step back and get out of the media cycle. AI is not saturating the business world, it is emerging. It impacts our lives in a variety of ways, but many times, we can still opt-out of the craziness and build an environment to get ahead without burning out.

You don't have to subscribe to hustle culture, work constantly, throw anyone under the bus, climb over the bodies of your coworkers, no.

THE HYPE MACHINE

AI is in every news cycle right now. "Grandfathers of AI" and "AI Founders" call for caution with AI because it might kill us all. In fact, the discourse around AI seems to take us to this Terminator-esque ending no matter what we do according to these people. Serial killers always get headlines and especially when they aren't human and plan on the total genocide of our species.

This type of hype and talk is a distraction at best and a blatant dismissal of the very REAL problems AI brings to our society today. Voices like the Algorithmic Justice League (https://www.ajl.org/), Dr. Joy Buolamwini, Timnet Gebru highlight real-world problems AI is creating today. By ignoring these problems and only thinking about the end of the world, we forget that someone denied a car loan because of algorithmic bias can lose their ability to get to work and then their job. No job, no money, no benefits, no healthcare. AI just destroyed this person's world.

Instead of the world ending, let's think about how we can break out of the hype machine and focus on doing our current jobs to build AI for Good. We don't need to wait for these "AI Luminaries" to fix the world. We can help in small ways every day.

WHAT IS AI'S REAL IMPACT ON YOUR JOB?

According to a recent Goldman Sachs report, "two-thirds of US occupations are exposed to some degree of automation" (https://www.goldmansachs.com/intelligence/pages/generative-ai-could-raise-global-gdp-by-7-percent.html).

Look at that statement, "exposed to some degree", that doesn't mean replaced. Automating your emails is "exposing" your job to automation. If you read email as part of your job and now your computer reads those emails to you using text-to-speech, that is automation.

In our context, **automation** is defined in Merriam-Webster as:

> *"The technique of making an apparatus, a process or a system operate automatically"*

With "automatic" defined as:

> *"Done or produced as if by machine"*

I've talked to people who believe their job cannot be automated because their role is the business of relationships. I've met people who are certain the robots are coming for their jobs because they're in the business of repetition. But the truth is...

SAYING THE QUIET PART OUT LOUD

If your company does not value your human contribution to your work, you are at risk of being automated away. Remember our discussion about Return on Investment (ROI)? If your cost to the company is greater than the cost of an automation doing the same work, at the same value, than the robot is going to get your job. Luckily, you can articulate your value as greater than that of a machine because you know your Core Competencies and you know you can grow skills more efficiently than any machine.

Automation is introduced into a process to reduce human involvement and either lower cost or increase value. Using Automation in a process can reduce errors, make the process incredibly predictable and extremely transparent. From an auditing perspective, automation is amazing (when implemented properly) to see what happened at every step of a process.

These savings from automation can be returned to the business as investment or stuffed into the pockets of the shareholders. Not all company cultures value reallocation of human capital (fancy way of saying redirecting humans to new work when their old work was automated). Fortunately, you will know these cultures when you see them. It is clear in their values which you will explore as you discuss your company's AI Ethics.

By getting in the driver seat of AI at your company, you can help steer some of these decisions or, at the very least, see early warning signs on where the company is going with their AI Journey. Leading the Red Team, engaging your Senior Leadership, taking part in the AI discussions at your company, all these activities will be the proverbial "canary in the mine shaft". Be engaged, be informed, and be active.

JOBS ARE CHANGING, COMPENSATION ISN'T

In the Age of AI, jobs will transform. I see this when I walk into McDonalds today. The manager once upon a time was accountable for their team to perform, handle customer complaints, and address employee issues. Now, you can add on top of this, Self-Service Kiosk Technician. When customers encounter issues with the kiosks, who's called to help: the manager. As more and more roles are converted to technology, more and more jobs will include an element of "digital babysitter".

Your job can follow similar parallels. If you work with an RPA bot that processes invoices, are you now responsible for managing that bot? Your new tasks (remember taskification):

- Monitor the bot's work to ensure accuracy
- Engage technical support if there is a problem
- Fix the bot's work when an error occurs
- Manage exceptions from the bot's workflow

In theory, the RPA bot should provide a cost savings on other tasks that offset these new tasks. The benefit to you is that you now can add "Digital Babysitter" to your resume. Future job markets will value this skill.

If the bot is creating more work for you, then notify your management team immediately and discuss a mitigation plan. Part of owning your career is going to be calling out the AI digital coworkers when they are not doing their job, just like you'd call out a human team member who is slacking. By not performing (human or digital), that team member is adding work to everyone else on the team. That's not fair or fiscally responsible. Engage in conversation to understand what is happening with the AI digital coworker and how that can be resolved with the help of your Red Team.

Don't say Digital Babysitter

I'm using that title in jest, but you should absolutely represent your work with AI on your resume. If you are managing and coordinating the work of an RPA bot, put it on your resume. I firmly believe your skills with digital coworkers will be a market differentiator soon, if not already. Whatever work that bot is doing, you are managing it if you are accountable for that bot's operation.

BASELINE YOUR NORMAL

What is normal? In your job, normal is the baseline measure that you use to track your performance. Normal should be comfortable but challenging, static but growing, and ambitious but achievable. This baseline will help you identify when something isn't quite right in your work life.

Your normal can obviously change as your job evolves with AI and automation entering into the process. As we discussed earlier, if there are sudden changes, ask "why?" As you evolve your normal with AI, document the evolution and talk about it in your one-on-ones with your manager. The key is to make sure that you have a visible connection between what used to be normal and what is now normal.

The taskification exercise is a great way to understand your normal. I'd encourage you to return to your task inventory and apply times to each task. How long does it take you to perform the "average" task?

I know, I know, "average" can be wide and varied, but you should have a general idea of how long something takes to do. If you're processing invoices, like the above example, you should be able to calculate how long an invoice takes to complete. If the invoice's data, such as type, submitter, division, etc. make the time to complete vary, that's okay. Track the average time for Type X Invoices and track it separate from Type Y Invoices. This will help you assess the value of automating a task.

Imagine Type X Invoices take 5 minutes per invoice. Type Y Invoices take 10 minutes per invoice. You might assume Type Y is the biggest opportunity for time savings. This is where you must understand volume and stability of the process for completing the task. If you have 100 Type Y Invoices a day but the process is highly variable because of complexities in the data, it might not be the better option to automate when 200 Type X Invoices come in a day but are extremely simple to process. Automation is best when you work with extremely stable processes, high volumes and high value tasks. But if you don't understand the "normal" or "average" effort that goes into a task, how can you identify the best options for AI to help you?

BEWARE THE HELPFUL

There are many people who try to be helpful, but because they are misinformed, they actually create problems.

I've seen this on countless automation projects where managers are directing how to perform the process to be automated. The managers explain, work with the team to document, and then when that process is shown to the people who actually do the work, the workers point out a bazillion things missing. The key is people who do the work know the work. Managers have good intentions but often are disconnected from the day-to-day and the realities of how workers get their jobs done. As managers, their job isn't to do the work but to manage those who do.

As a person who does the work, own your work and understand how it is done. That way, you can articulate to others how the work is done. This is the best way to be an authority on your work and be the best person to guide AI efforts to achieve cost savings and avoid project bloat due to rework (from having to fix all the manager's thoughts on how work is actually done).

AI WILL NOT IMPACT ME

I've met these people, and sometimes they are right. But beware being unaware. AI's capabilities are ever growing and tasks that we didn't think AI could do, "AI will never create a poem" or "AI will never create a piece of artwork" or even "AI can't read a document and understand what it means", these are not only AI enabled today, they are cheap for your average consumer to use. Technology marches on and what you think is impossible today becomes inevitable tomorrow.

Even if you believe your job is safe from AI, that is no reason to check out from the Age of AI. I argue that those jobs most "safe" from AI are the ones most augmented by AI. If one of your tasks is relationship management which breaks down into sub tasks like email communication, scheduling face-to-face touch-points, and responding quickly to VIP clients, AI can help you.

It can write engaging emails that drive your point in ways that you cannot craft on your own (depending on your writing skills). Meetings can be scheduled by an AI assistant as well as perform research about who you are meeting with, then write a report for you to consume prior to the meeting. Your VIP client calls can be transcribed in real-time for note taking so you can focus on the conversation and still receive the action items reported back to you or even scheduled automatically on your calendar. AI can help you even if your job is safe.

EXERCISE: EXPLORING AI TOOLS FOR YOUR TASKS

Now that we've discussed how AI can aid various parts of your work, let's make it more practical and relevant to you. This exercise aims to help you explore AI tools based on three tasks you perform at your day job. These tasks should be from your taskification exercise, conducted earlier in this book.

Choose Your Tasks: Select three tasks from your taskification list you believe might be simplified or automated using AI. This could include tasks related to data processing, email communication, scheduling, or any other task you think might benefit from AI augmentation. Remember your Core Competencies and AI's Core Competencies when selecting tasks to automate. Keep tasks that use your magic for yourself.

Research AI Tools: For each selected task, spend some time researching available AI tools that help automate or streamline that task. Remember, the goal isn't to eliminate your job, but to make your job simpler, giving you more time to focus on tasks that require your unique human touch.

Document Your Findings: In a notebook or a digital file, write the AI tools you've discovered and provide brief descriptions of what they do. Include notes about how these tools might improve the selected tasks. For instance, can an AI tool reduce the time spent on the task? Increase accuracy? What about negative impacts? Look for opportunities and threats to your existing work style.

Evaluate: Once you've documented your findings, evaluate the potential benefits of these tools. How would the integration of these AI tools change your normal? Make your job easier, more efficient, or perhaps even more enjoyable? What bias is brought to your work? Remember, AI tools always need to be seen through the lens of ethics, values, bias and culture.

The Dossier Dilemma

In preparation for a meeting once I had an AI system prepare a dossier on all the meeting participants by looking up their LinkedIn information. The idea was to have a brief report on each visitor that would show I did my homework on them. Human assistants often put these together for others but I don't have a human assistant so I used AI. The report was totally wrong about the attendees. It looked up the wrong people. Fortunately, I checked before the meeting and trashed the output before saying something I'd regret. Remember, always check AI's work.

Remember, the aim of this exercise is not only to familiarize you with AI tools relevant to your work, but also to encourage you to think about how incorporating AI into your daily tasks can redefine your normal in a positive way. It's about working smarter, not harder, in the Age of AI.

NO THANK YOU AI

You might not want AI to help you. That's okay too. You might think AI will detract from your autonomy or authority over the work that you do and you'd be right. Part of the value you get from AI is that it takes tasks off your plate. I am not advocating for AI in all cases. Sometimes it is appropriate and there are also times where AI isn't possible due to the complexity or cost of that task.

But if you are going to stay engaged, you need to understand when AI applies and how you can still own your work within the context of using AI to maximize output without overworking yourself. It's about having a sustainable career in an age where technology and automation are becoming the norm.

The Anti-AI stance isn't just for workers. There are businesses who are expressing no interest in AI. I've encountered many companies that "have been doing things the same way for 30 years and they're doing just fine" according to their founder. That's fine. Good for them.

Your competitors, whether as a business or individual (those other people who want your job), might invest in growing their AI capabilities and use that to differentiate themselves in the market. A common saying floating around LinkedIn is "You won't be replaced by AI. You'll be replaced by someone using AI." How true that is depends on the business or individual's ability to articulate their value, that magic we talked about early in the book.

Not that you should abandon your existing processes for some shiny new toy. However, it's important to remain open-minded and evaluate the impact of technology on your business before writing it off completely.

Whether you're a worker or a business, remaining curious and open to the idea of AI can help you stay relevant. Take time to explore how AI benefits your work, even if you're dreaming up out-there ideas. You never know what might come from those explorations.

WHAT TO DO ON TUESDAY

Let's keep this Tuesday normal. Your task is to reflect on your normal. What is normal for you? Why is it normal? When does normal get abnormal? Are their tools to "get back to normal" if things go off the tracks?

Timeblock: 30 minutes

Purpose: Reflect on your normal

People: You (keep this one to just you)

Process:

- Stay in your workspace. You will think about work better when you are in the context of the actual work, but turn off your notifications, set your Do Not Disturb on Slack/Teams/Zoom and focus.

- Complete this statement on a piece of paper: "My normal day is ________". Answer with whatever comes to you first. Whether that's a description of your day or a single word, the goal is to reflect on the definition of normal, not to fit your normal into a box.
- Complete this statement under the previous: "My normal is derailed when ___________"
- Complete this statement next: "My normal is restored when ___________" Sometimes normal cannot be restored after being derailed. If that's the case, write that. I'm thinking of doctors when a surgery runs long and derails their schedule for the rest of the day. There might not be the ability to recover other than to work a really long day. Also, if your normal is restored by stepping away from work and going for a run or eating lunch, or whatever, list that here.
- Complete this statement: "I wish my normal was ___________"
- Read your answers out loud to yourself. How do they make you feel?

Product: A written reflection on your normal and how it is impacted

Prep: Have some paper ready for your answers. Try to do this activity in your workspace if possible to get the feel of your work as you are reflecting on your normal.

Pitfalls: Be thoughtful about your answer. Don't cheat yourself by saying, "My normal day is sucky." Or "My normal day is so boring." If your answers are like this, you are in the wrong place, either mentally or occupationally. Perhaps these are signs of a necessary change.

Practical Concerns: Don't overthink it. Let the answers flow out of you.

USE CASE: AI, TELL ME ABOUT AI

AI can be very helpful in writing. I was working with a team that was writing updated job descriptions for existing roles. This company wanted to start with a baseline from industry standards, but they wanted to use their brand language and important facts about their company in the job postings. Using Jasper.ai, the company configured their brand voice and wrote the job descriptions. Then, working with their HR Team, the company tailored Jasper's output into job descriptions that fit the company's existing job code structure.

While this was happening, I encountered the Replaced by Robot website (https://www.replacedbyrobot.info) and challenged the HR team to consider how AI might augment these roles. In a tight labor market, we sought to use AI as a differentiator to tell candidates, "here, you will use AI to grow your career for the future". We could have used the exercise in this book to investigate AI opportunities per task, per job, but that would be a lot of searching. Instead, we asked Jasper to help identify opportunities for AI to enhance the role. Our prompt was:

Given the following job description, how would you recommend we use AI to perform the tasks of this job more efficiently: [Job Description]

This led to a list of potential AI opportunities for each job. Working with the HR and IT team, we vetted the list for the jobs and removed items we knew would not work in the company because of technological, security, or compliance issues.

AI tools like LLMs can tell you a lot about AI. Always remember to vet the output to avoid hallucinations and factually inaccurate content. Fact checking is a critical skill in the 21st century (it always was, but more so now than ever before). Use LLMs for idea generation and how you can improve a process, task, or even a job with AI.

CHAPTER 8: PROBLEM PATTERNS

In the world of software development, we have a concept called "Design Patterns". These are blueprints or recipes for building software functionality. When a design problem is encountered, such as how to improve performance by making a code object clone itself instead of creating a whole new version (Prototype Pattern), Design Patterns are used as a tool to implement a known, tested, solution.

For this chapter, I'm going to give you a bunch of problems and AI solutions, along with a few of the right questions to ask about the AI solution. While this list is not encompassing all business problems, I hope that you'll see some of your problems in those that are provided.

Each Problem Pattern will follow a standard design:

- **Title:** What is the problem we're solving
- **GripeChat:** Users are chatting to each other complaining about some issue in their business.
- **Problem Statement:** We take the GripeChat and show how to convert it into a problem statement.
- **AI Solutioning:** Using the problem statement, we

consider how we could solve it with AI (or not).

- **Problem problems:** What problems are created by the AI Solution.

Use this section to spark ideas on how you could approach one of your problems with AI.

WORKING THROUGH A QUEUE

Many knowledge work jobs have work queues. In this problem pattern, we're going to do the same action repeatedly on a list of work so the human can focus on the exceptions in the process instead of the tedious tasks.

GRIPECHAT

Aaron: Ugh! If I have to process another invoice for ABC Corp, I'm going to lose my mind!

Becky: Awfully dramatic. Isn't that what you do all day?

Aaron: Yeah, well, my brain is melting from doing the same thing over and over again. Want to switch accounts a while? I'll take XYZ Corp and you take ABC?

Becky: Sure, just don't forget to do the extra steps with XYZ. Don't want your melty brain screwing up my client. I'm dealing with a few problems already, don't want you adding to it. But I could use your help. At this rate, there's no way I'm getting these done by the end of the month. The problem invoices are slowing me down.

Aaron: They're pretty much the same. Just gotta click the extra buttons for XYZ. I think my brain will survive. Thanks for the switch! I'll get you caught up. ABC is pretty straightforward, and I haven't hit any issues all day. If there's a problem, just let me know.

PROBLEM STATEMENT

Maybe Aaron just needs a break, but anytime someone identifies that they are doing the same thing repeatedly, with little variation to the task and a high volume of work, then you have a business problem. The problem here is performing this work has a cost. With invoices, there is an operational cost to processing those invoices and, if the invoices are not processed in a timely manner, the company's cash flow will be impacted. Fortunately, this task is a prime candidate for AI.

But first we need to articulate the problem.

- Who: Invoice processing team
- What: Are struggling with processing their invoices
- Why: Because the monotony of the work and the delay caused by errors
- When: Throughout the month, but especially during crunch time (closing out the month)

Our problem statement is:

The invoice processing team is struggling to process the invoices for ABC Corp and XYZ Corp because of the monotony of the work and exceptions throughout the month. The problem is amplified during month's end.

But how much does this problem cost the company? To

calculate this, we examine how many ABC Corp invoices come in a day. To keep the math simple, we'll say 80 come in a day (or 10 every hour). Aaron works 8 hours a day, and it takes all 8 hours to process the 10 invoices (we're keeping the math simple here). If we take Aaron's cost to the company, let's say he makes $19/hr, then we can calculate that 1 invoice costs the company $1.90.

AI SOLUTION

Before going any further, answer the question:

Should we use AI to solve this?

Let's use the AI Ethics Wheel to examine the ethical impacts of using AI in this problem.

- **Human Rights:** an automation to help process the invoices wouldn't impact human rights but it could impact people's jobs. We need to determine if the savings generated by automating this task will apply to reskilling the impacted team members. After all, if the AI can process the invoices 24/7, do we need the same number of Invoice Processors to complete the work?
- **Bias:** We know ABC Corp's data and process are very standard and so to keep our initial AI solution simple, we will focus there. This means that we are implementing a Selection Bias in that we'll be focused on ABC Corp data. To ensure we avoid Decision Bias, we will work with Aaron to understand the existing process, so there is no confusion on how the current process is performed.

- **Inclusion:** This automation will make Aaron's processing metrics look amazing compared to Becky's. Managers will need to update their performance expectations accordingly to avoid penalizing Becky and unduly rewarding Aaron.
- **Privacy:** The AI will not store data about ABC Corp's invoices but will use a 3rd party computer vision system. That means the invoice data will need to be transmitted for processing. We need a vendor we can trust and to ensure that vendor is not storing the data provided.
- **Explainable AI:** This AI will focus on process automation and computer vision. We need to ensure clear documentation on the process and that the vendor used for computer vision (reading the invoices) has robust documentation for their product.
- **Level of Autonomy:** This automation could run on its own, but to ensure human oversight, we will build in an element of human review to the work. This way, the human can intercede if needed.

For the components of the AI Solution, I see a process automation system and a computer vision system needed. The process automation will be Robotic Process Automation (RPA) because the application Aaron and Becky use to process invoices does not have any APIs.

An API is a bridge that software developers use to connect applications together. Using APIs, software systems like Microsoft Word can read data from Microsoft Excel. If the software system does not have an API, then you need to interact with the data through the application's user interface. RPA can be taught to interact with that interface just like a human does (scroll here, type in that text box, click the save button).

Another component we'll need is a computer vision capability. The invoice files are scanned documents that Aaron and Becky have to open, review, then copy and paste the data from the invoice document into the invoice processing application. Many RPA tools have computer vision tools built in or accessible to them. For this project, we decide to use Microsoft Azure Cognitive Services for computer vision because it has thorough documentation and this company already uses Microsoft Azure for other IT projects. But, before finalizing the decision on vendor, the IT team would compare pricing of the computer vision system along with the cost of the RPA license and the cost to develop the solution to ensure that the return is greater than the investment.

Back to math...let's say the RPA license is $5,000/year. The computer vision charge is $0.25 per transaction (reading 1 invoice = 1 transaction). This math gets confusing because it is on different scales. Aaron's time is by the hour. Invoice counts are by the day. RPA license is by the year. Computer vision is by the transaction. To make sense of it all, you need to go to the common denominator, which, in this case, is the year. We know Aaron processes about 80 invoices a day. We know Aaron works all year. With vacation and holidays, Aaron works 47 weeks a year.

47 weeks at 40 hours a week processing 10 invoices an hour at a cost of $1.90 per invoice (47*40*10 * 1.9). The ABC Corp invoices cost Aaron's company $35,720 a year.

Now we need to calculate the cost of the solution. Let's say it is $5,000 for internal development, plus the RPA license, plus the cost of all the computer vision transactions (5,000 + 5,000 + (47*40*10*.25)). This results in a cost of $4,700 a year (assuming the $5,000 development cost is continuous to keep the system up to date with process changes). Great news! The cost of $4,700 is less than the generated value of $35,720. But then reality comes crashing down and we realize that only 70% of the invoices from ABC Corp can be processed without exception. This lowers the generated value because not all those invoices can be processed by the automation. Here's the new equation: 47 weeks * 40 hours * 7 invoices * $1.90, is now $25,004 in value generation. The cost per invoice should go up, but Aaron doesn't do a good job tracking how long these exceptions take to complete. He doesn't know how the exceptions skew his numbers because he's never paid attention to that before.

We are still in a good return on investment, but you can see how calculations can skew quickly.

PROBLEM PROBLEMS

- Make sure you clearly understand the business process as performed by the person doing the work
- Ensure you have a dataset that represents the data you will work with. Test not only against the most orderly and clean invoices in the database. Make sure your

data represents the real data flowing through your system. Use a wide range of invoices as the training set for your computer vision model.

- Build a plan for what Aaron is going to do after the automation is implemented. If this automation is performing 70% of his work, what is he doing with that 70% of time now? Redeploy his skills or train him to be Becky's support team member to help with XYZ Corp during this newfound free time.
- Ensure the RPA bot is reporting its work progress so you can ensure the ROI calculation is accurate. What if the bot is only processing 40% of the invoices due to an invoice change? What if the flow of invoices triples? Use reporting and monitoring to ensure you keep a close eye on your digital workforce (in this case, the RPA bot).
- Don't forget to adjust performance expectations for both Becky and Aaron with this new AI helper. Is Aaron managing the bot's work? Who does the bot report to?
- What is the plan to keep up with the application and process changes for the RPA bot? Processes evolve, applications change, bots will need to be retrained to keep up. Make a plan and stick to it so you can stay ahead of any RPA outages.
- Is your computer vision system designed for invoices? Certain computer vision systems with low-code have a document comprehension component, which enables you to ingest invoice formats quickly. Just beware that

if your invoice changes, if the client makes a format change without telling you, your computer vision system might need to be retrained.

PREDICTING PROBLEMS PRIOR To PANIC

You want to tell the future? Predicting business conditions is something every executive wants. Whether you can do it depends heavily on your data. In this supervised learning, regression problem we will be predicting machine maintenance.

GRIPECHAT

Yuri: Oh man, another shut down. What the heck. This means weekend work!

Zeek: Ugh, yeah. The fan on unit 2 blew again. I told you it was rattling too much. Told the techs too, but do they listen to us linemen? No.

Yuri: We've gotta hit 200 units by Monday. This weekend's going to suck.

Zeek: Well, maybe they'll fix it quick. Unit 2 is supposed to be that internet connected one with all the computer stuff in it. Maybe it's a quick fix.

Yuri: Is it ever?

PROBLEM STATEMENT

Unexpected outages cost manufacturers an estimated $50 billion each year (https://www2.deloitte.com/content/dam/Deloitte/us/Documents/process-and-operations/us-cons-predictive-maintenance.pdf). This is a major problem for obvious financial reasons but also a huge impact on the workforce. As we can see here, Yuri is upset about weekend work. This can lead to employees leaving a company for more stable work conditions. This employee churn is extremely expensive as well, with some positions costing up to 200% their annual salary to be replaced (https://lrshrm.shrm.org/blog/2017/10/essential-elements-employee-retention).

Let's articulate the problem:

- Who: Line workers
- What: Are being disrupted by the machines shutting down unexpectedly
- Why: Because the equipment isn't working as expected
- When: During operational hours

Our problem statement is:

The line workers must work weekends because the machines are unexpectedly breaking down during the workday.

When we think about the cost for this problem, we have the cost of lost manufacturing time, compounded with the cost of employee retention. Other costs to consider here are the actual costs necessary to fix the broken machine, among others. Problems with the machines in manufacturing spiral out to other problems down the supply chain either within the company or throughout the distribution channels. An unplanned outage is a very costly affair.

AI SOLUTION

As always before going any further, answer the question:

Should we use AI to solve this?

Let's use the AI Ethics Wheel to examine the ethical impacts of using AI in this problem.

- **Human Rights:** probably no impact here. Predicting when a machine is going to break is a good thing to ensure people are safe.
- **Bias:** Do we have data about this machine? In the story above, we know this unit is internet connected and probably has a lot of data, but if no one ever baselines that data, or the data was not stored, you will be missing important data to act on.
- **Inclusion:** Predictive maintenance can impact your repair team's schedules. Do you have repair windows during their normal work hours or during off hours? Remember, the line workers are not the only people impacted by this AI system.
- **Privacy:** We will imagine our system is 100% internal to the company. Telemetric data (data about a system's performance and usage) is often not identifiable to a person but always make sure. Is the user's information stored in the telemetric data? If so, create plans and procedures to scrub it out. You don't want personally identifiable information (PII) mixed in with your systems telemetry data, as it is often not secured to the degree that PII should be.

- **Explainable AI:** You are going to want data scientists for this one to ensure the prediction process is vetted, examined, tested, and validated. While there are Low-code tools that work as prediction machines, for a problem this big, you'll want to have "the one throat to choke" when things go awry.
- **Level of Autonomy:** This isn't a concern because the system will alert the repair team and nothing else.

For the components of the AI Solution, there is only one AI element needed: the prediction machine.

To start, inventory the data sources that you'll need for the project. You need the device's (Unit 2's) data. Let's imagine the data is stored in a log file for each day on a cloud server. The log file is a long list of data points and three of those points are informative of an impending failure.

Now, we need to make this log file into a structure dataset so we can more easily work with the values. Working with our IT team and using data transformation tools, the IT team converts the log file into a spreadsheet format with the following columns:

- Timestamp
- Fan Amplitude
- Fan Frequency
- Fan Accelerometer Speed

Each row of the spreadsheet would be an entry from the log file. Each log file tracks the status of the fan every minute (we'll keep it simple here) and records that to the file. Our spreadsheet is compiled every four hours by transforming each entry from each minute into a row with these four values. For one day, we will have 720 entries assuming a 12-hour plant operation.

With the spreadsheet of data, we would now use a tool like Orange Data Mining (https://www.orangedatamining.com) to explore the data and search for correlations. Is there a link between these fields to signal an impending failure? By examining the data, we can see a relationship between the frequency increasing and the accelerometer speed decreasing. Using this data, we can build a regression problem to predict the likelihood of failure within the next 72 hours. Using all the historical data Unit 2 has collected, along with data from the manufacturing to signal optimal operation specs for the Fan values, we can build a model to show the percent likelihood of failure in 72 hours. As that percentage increases, we can engage our repair team with the urgency to avoid an unplanned failure.

As more data comes from other units, we keep optimizing the prediction system to get more and more accurate predictions. This doesn't mean that Unit 2 will never break down unexpectedly, but we will reduce the volume of such events and help our line workers keep their weekends for themselves.

PROBLEM PROBLEMS

- Do you have the right information to predict the outcome you are trying to predict? What if Unit 2 had great data but Unit 3 did not and Unit 3 was a different type of machine? Your prediction system would not work on Unit 3. Make sure you have data that represents the population impacted by the AI system.

- Think holistically about the impact of your prediction system. What is the impact on the line workers? Repair team? Management? Vendors? Service contracts? Could you see an increase in service costs because you never use your service hours and thus your service vendor raises your pricing to make up the difference (fixed fee vs. hourly work costs)?
- How are you constantly care and feeding your prediction system to ensure it is learning from the most accurate data? Plan that your AI system will require maintenance. AI is not a software that you build once and let it run forever.
- Did you select the right data to answer your problem? Here we're using fan data, but what if the shutdown was from temperature? Is the fan still the problem? No matter what you are trying to predict, you must explore the data to discover the right features that impact the outcome. Remember, features you did not expect to impact the outcome could have a major role. Don't give in to confirmation bias and ignore the facts based on your beliefs.

WHAT MATTERS HERE?

Using a Large Language Model (LLM) and generative AI, we'll explore summarizing a document to focus on what matters to the analyst.

GRIPECHAT

Gretta: So, I just got the customer analysis report from Kyle. Did you see it yet?

Henry: Yeah, 150 pages! That isn't a research *brief* that's a research *dissertation.*

Gretta: It's due Friday. I don't think I can plow through this dense content in three days.

Henry: Maybe we can get a TL; DR; from Kyle, if we ask nicely?

Gretta: Yeah, that might be enough to send the VP our analysis. But what if we miss something important?

Henry: You read half? I read half?

Gretta: Same issue. Maybe there's a connection between the halves we need to know.

Henry: Ugh, Kyle!

PROBLEM STATEMENT

That's a long brief. In this example, Gretta, and Henry have to provide a report to their VP. Their report needs to summarize Kyle's content without missing anything important.

Let's articulate the problem:

- Who: Research analysts
- What: Need to summarize a long report
- Why: Because their VP needs to know what's important in the report
- When: In the next 3 days

Our problem statement is:

The research analysts need to summarize a 150-page report for their VP in the next three days.

I'm going to skip costs on this one but encourage you to explore the time it would take Gretta and Henry to read, summarize, and prepare a presentation of the content. This would not be a few hours. Dense research content can take many readings to grasp and many more to understand.

AI SOLUTION

As always before going any further, answer the question:

Should we use AI to solve this?

Based on this book, we know a LLM is an excellent tool to use for this summarization task. Let's use the AI Ethics Wheel to examine the ethical impacts of using AI in this problem.

- **Human Rights:** when dealing with customer data, depending on what you're selling, you are most likely not dealing with Human Rights issues. In this fictional scenario, we are clear of human rights issues.
- **Bias:** When summarizing content, the LLM will decide what's important and what is not. Is the LLM specially tuned to specific types of data? If the LLM is "instruction tuned" on a specific topic, say "Marketing Content" or "Environmental Impacts", it could create biased output as the LLM leans into those topics. Be aware of any special/instruction tuning your LLM has,

which will form bias in your output.

- **Inclusion:** Many modern LLMs are trained on the data of the Western world (North America, Europe). If your customer report is based on a customer segment in South Central Asia, your LLM might not have the appropriate data for this community. If your summarization skews to the wrong audience, your output could be misleading or just plain wrong.
- **Privacy:** If the LLM runs in-house, the IT team has set up the LLM in a private environment for this company, then privacy might not be a concern. If you are using a Black Box AI (3rd Party system like OpenAI's ChatGPT), you will expose the data in the client report to OpenAI. If this is acceptable based on your AI Red Team's rules of the road, then you are fine, but understand what data is being exposed to who and how they can use it in the future.
- **Explainable AI:** LLMs are very difficult to explain how they make the decisions they make. Their large nature with billions of features makes tracing why the output was generated very challenging. Many LLMs are building the ability to cite sources as part of their features, and this would be very important for this summarization activity. Here, Gretta and Henry would want to see where in the client report the summarized data is coming from. This can be done with many recent LLM implementations, whether it is a tool like ChatGPT, Jasper or a private LLM like Llama and Microsoft Azure OpenAI Services.

- **Level of Autonomy:** There is not a concern over autonomy because the output must be finalized by Gretta and Henry. The key challenge here is ensuring that the team using the LLM understands that they must treat the AI's output as a first draft. LLMs do not produce final content but can get you started down the road. The team members using the LLM are responsible for the output.

I started this section setting the expectation that we would use an LLM to solve the summarization problem. LLMs are enterprise tools and not deployed to fix a single problem. In this situation, the AI Red Team would provide guidance on how to use the company approved LLM for summarization tasks. That can be a policy document, such as:

AI RED TEAM GUIDANCE ON LLM APPROVED USE CASES

Policy: The approved LLM may be used for document summarization purposes on internal business documents.

Standards:

- The approved LLM is the company's internal instance named Hawkeye (this is a private cloud deployed instance of Falcon 40B).
- All documents to be summarized must be stored in the approved storage environment for their security level. See Secure Document Storage Policy for more information (I didn't include this here in the example).
- All output must be reviewed and approved by a human resource who is accountable for the output.

- Users must never attempt to alter the LLM's behaviors or personality in the output through their prompts.

Procedures:

- Upload the document to be summarized into the appropriate storage system.
- In Hawkeye, refresh the search index (imagine this is a known capability of the Hawkeye system)
- Enter your prompt for summarization and get the output from Hawkeye.
- Export the output into your destination document.
- Review the output to ensure accuracy and the values of Company M (our fictional company) are demonstrated throughout the output.

Non-adherence:

- Using an LLM that is not Hawkeye will lead to disciplinary actions commensurate with our employee agreement policies
- Consequences for storing the document to be summarized in a manner that is not compliant with the Secure Document Storage Policy are covered in that policy.

THE TECHNICAL COMPONENTS

There are a few components of this LLM summarization solution.

The first is the LLM. Running a Large Language Model is not a small endeavor. These systems (the higher functioning ones) take significant resources to perform their functions and a robust security infrastructure to ensure only those should have access can have access. There are a few decisions to make when selecting an LLM.

- 3rd Party Black Box AI: Do you want to use a 3rd party to host and run your LLM in their environment? This is easiest from a management perspective, but for that savings, you lose control of your data and the platform.
- Private AI: Do you want to manage the LLM platform? This allows you the most control, privacy and security, but comes with the management price tag along with another decision:
    - Local: Do you run the LLM in your company's data centers? This has a significant increase in costs as you will need to manage the hardware and software along with networks running the LLM.
    - Cloud: Do you run the LLM in a cloud environment like Microsoft Azure or Amazon Web Services (AWS)? These cloud environments allow the best of both worlds between 3rd party and private AI. You maintain control (depending on your configuration) and still get the benefit of others ensuring the infrastructure of your deployment. Always remember, no one cares about your data or security as much as you do. Treat it with the care it deserves.

In this solution, we will assume Company M has opted to use Private AI. This also means Company M's IT team has to identify a data storage location for the documents to be summarized. In our fictional example, we'll imagine the IT team is using a cloud environment and thus uses cloud blob storage (blob is a large, unstructured data format perfect for documents and images).

The storage and the LLM are just the backend systems to make the solution work, but there still needs to be a place for Gretta and Henry to interact with the LLM. In our policy document above, there was mention of "Hawkeye". In our story, we'll imagine Hawkeye is the web application Gretta and Henry will use to interact with the LLM. Users do not interact directly with the LLM, there is normally some kind of interface layered on top of the LLM that is user friendly.

As an example, ChatGPT is the interface. The LLM being used is GPT3.5Turbo.

IT deploys the LLM to process the summarization requests. IT deploys a web application, called Hawkeye, to receive the requests from the users and display the output. And finally, IT deploys the document storage solution to hold the documents the LLM will read and summarize.

Now, Gretta and Henry have to summit their prompt to Hawkeye. What do they write? Follow the [WHO] [DO] [WHAT] [HOW] format discussed earlier in the book to get the following example:

> *You are the VP of Marketing for Company M. Summarize the Client Report to highlight the items that are most important to me in a bullet list. Cite where in the document you found this*

information. Create a second list that displays items you did not highlight and explain why you think those are not as important.

Always ask your LLM to justify its decisions. Remember, it is a smart intern, not an expert. The system must justify and explain its decisions to build trust in you the user. Do not blindly trust the AI without proof it knows what it is doing.

PROBLEM PROBLEMS

- Have you trained your team on how to work with an LLM using a skeptical eye? Build strong fact checking muscles and editorial skills as the LLM's output will need review. AI does not produce a final draft.
- Monitor your cloud spend on the solution to ensure you are not surprised with a large bill at the end of the month. LLM infrastructure can be expensive, with each transaction increasing the cost (depending on the implementation). This solution supposedly reduces cost by saving time. Track the utilization of the LLM to ensure it is producing the value you expect.
- Beware of how your data is being used in 3rd party AI platforms. I am not saying do not use them. There are many use cases where a 3rd party, specially tuned LLM, is a perfect solution. I'm encouraging you to be an educated consumer and understand what data is being entered into the system and what that means for your company. Remember, once data leaves your environment, if you do not have a clear license around permis-

sive use, you have lost control.

- Provide guidance on how to build prompts for the team using the LLM. A lot of the quality of the LLM's output will be wrapped up in the LLM's training and the prompt. Without guidance on how to build prompts for the task the team members are trying to achieve, you will not get the savings you hope to achieve.
- Many open source LLMs have a Model Card. Using these Model Cards, you can understand the LLM (or any AI model that has a card) better in topics such as recommended usage, environmental impacts, known bias, and many other factors. Margaret Mitchell, Simone Wu, Andrew Zaldivar, Parker Barnes, Lucy Vasserman, Ben Hutchinson, Elena Spitzer, Inioluwa Deborah Raji, and Timnit Gebru provided an excellent paper on the importance and method of documenting AI models to ensure transparency. Their paper can be found at https://arxiv.org/pdf/1810.03993.pdf. Websites like HuggingFace list Model Card information in the model description when provided. Use these model cards to help you understand the AI model and be weary of AI systems without a card or with a card that was only half-heartedly populated. See an example Model Card for Llama 2 here: https://huggingface.co/meta-llama/Llama-2-7b
- There is a growing attack method known as Prompt Injection. Using Prompt Injection, users can make the LLM go around controls that you might have in

place, such as defining the LLM's personality or setting guardrails against types of content the LLM should never discuss. Monitor the prompts being used in your system to ensure the users are playing by the policy rules.

VANISHING CUSTOMERS

In the world of business, it's crucial to keep tabs on customer engagement. But what happens when customers vanish? This problem pattern will help you identify and address vanishing customers, ensuring you can stay connected and provide top-notch service.

GRIPECHAT:

Marco: Ugh, I don't get it. We put so much effort into our marketing campaigns, but our customers just disappear into a black hole before checkout!

Nicole: Tell me about it! It's like we're hosting a hide-and-seek competition, except our customers are the masters of hiding.

Marco: Exactly! We need some kind of magical tracking device to find them and bring them back into the fold.

Nicole: If only we could wave a wand and make them reappear. We'd be customer engagement wizards!

PROBLEM STATEMENT:

- Who: Customer engagement team
- What: Customers mysteriously becoming unresponsive

or disengaged

- Why: Lack of visibility into customer behavior and preferences
- When: During the purchasing process

Let's formulate our Problem Statement:

The customer engagement team cannot understand why customers are disappearing due to lack of visibility into the customer's behavior during the purchasing process.

The team at Company M is struggling with customers who seem to vanish without a trace. Despite their dedicated marketing efforts, they find it challenging to track customer engagement, leaving them in the dark about customer behavior and preferences. This lack of visibility hampers their ability to deliver personalized experiences and maintain strong relationships with their customers.

AI SOLUTION:

To tackle this problem, an AI-powered solution can come to the rescue. By leveraging advanced technologies, such as machine learning and data analytics, businesses can gain valuable insights into customer behavior, predict engagement patterns, and proactively address potential issues.

As always, should we use AI? Let's look at the ethics wheel:

- **Human Rights**: Our ecommerce site will not create issues in human rights but if you are a non-profit serving marginalized groups who could be at risk when using your services, be mindful that your AI systems will handle sensitive data.

- **Bias**: What is your historical customer data? What populations are illustrated in that data? Identify the biases in your data by asking the questions provided in previous chapters about data bias. While your data is based on your history, don't forget that your future customers might not be represented. Plan accordingly.
- **Inclusion**: Don't forget your loyal customer's behaviors while trying to surface the behaviors of customers who abandon you. Often when humans focus on solving one problem, we only see the problem and not the good things surrounding that problem. Beware setting up rewards/enticements for customers abandoning you at the expense of supporting your loyal customers.
- **Privacy**: Be clear to your customers what you are tracking and why. While you might think, "It's in my terms and conditions and my privacy policy, people will get it", consider when the last time you read one of those documents was and ask yourself, do you really think your customers behave any differently? Go the extra mile to tell your customers what you're doing and avoid any strife in a future discovery of how you've used their data to fuel your AI system. Also, be clear on how to opt-out in their language. Test your language to ensure it is understood by your customers.
- **Explainable AI:** Using machine learning and our robust data, we can create the data models that explore and explain why a customer is identified as at risk of

abandoning their cart during checkout.

- **Level of Autonomy:** Make it easy for people to opt-out. I firmly believe this will be a future differentiator in the market where companies who put the customer in the driver's seat to control their own data will win.

The AI solution employs machine learning models trained on relevant customer data, such as past interactions, purchase history and demographic information. With proper data collection processes in place, it's crucial to ensure the data is comprehensive and representative of the customer base. Regular validation of the data helps to maintain its accuracy and reliability.

Let's start with data collection. In this solution, we'll need customer data from our ecommerce system and probably our CRM (customer relationship management system). These are both structured data sources that we use to export a spreadsheet of a few tables:

- Customer information: demographic information that was provided during the checkout process
- Customer behaviors: such as what products they visited, which they put in their cart. This data started prior to the customer identifying who they were and used a tracking pixel to identify the customer. Once the customer hit a specific step in the checkout, we link the pixel with their customer record.
- Order information: what is in their shopping cart when they enter the checkout process

With all this data at our fingertips, we explore the data using tools like Orange Data Mining. Through examining the data, we discover a few interesting data points by using Clustering (Unsupervised Learning) to see related data records. We see that carts that were abandoned contained items with a price point over $500 and a customer's behavior of adding and removing that item over two times. They were indecisive about if they wanted to buy that item.

With this information, we remove the customer demographic data because it was not important in our clustering data. Customer behaviors and Order information mattered, so we only use that information going forward. Based on the historical data, we create a column in our data called "At Risk" and manually label (Yes or No) the orders at risk based on the demonstrated customer behavior of adding and removing the items from their cart when the items are greater than $500. We have now created the data for a Supervised Learning Classification problem.

Split the data at this point so you can train a supervised learning model on data that is labeled. Using this labeled data allows you to teach the machine learning model the right answer. With your trained model, you can now test the data you split out from the training set. Make a copy of that data (the testing data) and remove the answer from the "At Risk" column. Test your AI Classification model to see if it accurately predicts the outcomes and document how accurate the model is. This baseline is critical as you focus on improving the model and tracking progress.

PROBLEM PROBLEMS:

While the AI solution offers great potential, there are some potential risks and problems to keep in mind:

- Overreliance: Depending solely on AI may lead to overlooking valuable human insights and subjective considerations. If you used AI to mine out the reason for abandoning the cart, did you overlook anything like poor user experience in the ecommerce platform? Sometimes the answer isn't in the data and your selection bias (you didn't look at app utilization data) can mask the real problem.
- Data Quality: Inaccurate or incomplete data can impact the reliability and effectiveness of the AI system. Your customers might have all opted-out of your system or opted-out in such quantities that your calculations are not useful. This is a challenge you must navigate.
- Unintended Consequences: In rare cases, the AI solution may generate unexpected outcomes or unintended biases, requiring careful monitoring and adjustments. What if the system flagged someone as an at-risk customer incorrectly? What happens? How is the customer's experience impacted?

By considering these factors and addressing potential challenges, businesses can use AI technologies to enhance their understanding of customer engagement, improve communication, and bring those elusive customers back into the spotlight.

Now, armed with this pattern, you're ready to track down those missing customers and turn them into loyal advocates of your business!

UNLEASHING THE CREATIVE BEAST

Ever hit writer's block? Can't think of how to approach a creative problem? Don't worry, AI can help.

GRIPECHAT:

Sally: How's the ad campaign going?

Reeba: Don't remind me. Nothing's coming. I can't think of a way to tell our internal staff about our new products. They never care.

Sally: They care, we just need to speak their language. Maybe if we put a kitten on it? Everyone loves kittens?

Reeba: We tried that last time. Remember the angry emails from the dog lovers?

Sally: Okay, so puppy this time?

Reeba: But what do we say? I can't figure it out.

PROBLEM STATEMENT:

- Who: Frustrated creatives
- What: Cannot come up with a way to communicate the product changes in a manner the internal team will engage with/care about
- Why: Their failed previous attempts have created writer's/creative's block
- When: They are trying to communicate this new product internally

Here's the concise problem statement:

> *The creative team is having a challenge figuring out how to communicate the new product launch in a way the team will engage with because of the pressure of timelines and previous lack of engagement.*

Many of us have experienced the grueling battle with a creative block. It's as if our creative mojo has vanished, leaving us staring at blank canvases or struggling to find innovative solutions. The lack of inspiration, mental fatigue or self-doubt can be overwhelming, hindering our ability to tap into our creative potential.

AI SOLUTION:

Fear not, an AI-powered solution is here to help you conquer the creative block! By harnessing the power of LLMs, you can generate ideas and test them with your audience to see where engagement soars and sinks.

But let's not forget to evaluate the ethical considerations associated with this solution:

- **Human Rights:** Freedom of expression is a key component here. Ensure your team knows the AI is not the master of their creativity, but a tool. The team member's expression and creativity is a value and should be treated as a value-add, not as a cost.
- **Bias:** As stated in previous examples, depending on the LLM's training, the output could be biased and

restricted. Understand the bias built into your LLM and plan to mitigate it. For creative projects, I encourage a diverse team collaborating with the AI system to produce the output and not limiting yourself, or your company, to a mono-culture.

- **Inclusion:** Every artist deserves a chance to shine. Encourage your team members to be an active participant with AI in the creative process. Do not set the AI's time savings above your team's creative value. Set practices and standards of how creatives add to AI output to communicate you value their work.
- **Privacy:** In a business context, the creative work probably belongs to the company and if you are using a private LLM, this should be fine. Be conscious of a 3rd party LLM and whether that should have access to your creative works. Also, if your creative brief (the content you share with the LLM to build out the creative output) contains sensitive information, implement whatever security procedures are necessary in your organization. If you are making a financial performance presentation with private information, be thoughtful about your LLM use. I would encourage not using the LLM for that section of the presentation.
- **Explainable AI**: As stated earlier, LLMs are complex and often difficult to explain how it came to its conclusions. Be aware of any special tuning or instructions that have been provided to the LLM such as a focus on building presentations (like beatuify.ai) or marketing content (like Jasper).

- **Level of Autonomy**: Ensure your team understands the LLM is a contributor, not the director. The team owns the work, not the AI.

Earlier, I talked about implementing an LLM solution. See "What Matters Here?" for details. In this solution, I will highlight creative works with AI and LLMs.

To start, your creative team needs to talk to their audience and understand what content they would like to see. As in every marketing effort, know your audience. In the early chapters of this book, I shared how to understand "Who", in this case who = audience. I encourage using Design Thinking tools from designkit.org to understand your audience. For ease of reference, here are a few questions to ask your target audience for your next AI creative endeavor:

- What was the last memorable commercial you saw?
- What communication mediums work best for you? Text? Video? Posters? Audio?
- Who is your go-to person to learn about new things at work?
- When did you hear about the last product update? Did you care?

Those questions are just the tip of the iceberg. You can use your Design Thinking skills to build empathy with your audience and discover how to communicate. For us, we're going to use the answers to these questions to springboard into some prompts for the LLM.

When the audience tells you the last memorable commercial

they saw was the one with the inflatable monkey crashing into a car, here's a prompt to use:

> *You are a product marketing specialist and need to announce new features for our product line. My audience likes a commercial where a giant gorilla is holding a sign about buying car insurance as it crashes into a brand new car. List ten ideas for our new product [describe the product here] using this commercial concept.*

You can see how we used the interest in the commercial as an input to add context to the prompt. Your research into your audience should feed your prompts.

If you don't know your audience and can't ask them, you can ask the LLM. Always validate the answers the LLM gives you with your real-world audience population. Never assume the LLM is right. Here's a sample prompt you can use:

> *You are a product marketing specialist and need to announce new features for our product line. List the top 5 care abouts I should address for our audience who is [information about your audience]. Highlight how [features] will benefit them.*

Ask follow-up questions to expand or direct the ideas (remember Chain of Thought Prompts?). Don't let the interaction with the LLM be a one-and-done interaction. As an example, let's imagine one of the 5 answers provided is "Mention how the audience will have more free time to enjoy with their family due to [feature]". Pull on that thread with the next question to the LLM:

> *Give me two ways the audience would resist the idea of our product, giving them more time with their family.*

Work through each idea provided, seeking the benefits and points of resistance your audience could have. Remember, LLMs are reason machines as much as they are content generators. Use them to help you think through your problem (audience engagement). Do not stop at single questions. Turn your creative process into a conversation with your peers and your LLM. Everyone should contribute, question, and collaborate to discover the best solution to your creator's block.

PROBLEM PROBLEMS:

While the AI solution ignites the fire of creativity, it's essential to be aware of potential risks and problems.

- **Dependency:** Relying solely on AI-generated ideas may lead to a lack of originality or homogeneity in creative output. These systems are not designed to create novel ideas but to stitch together the statistically most likely next word. Humans can turn that thread into a creative idea. Never underestimate the creativity of humans, especially those who claim to not be creative.
- **Algorithmic Limitations:** The AI system's suggestions are based on existing data, which may constrain

the generation of truly groundbreaking ideas. Similar to the above, but I'd also encourage you to understand that the LLM is not a fact machine. It is a word machine, a reasoning machine, and it doesn't know everything. It will act like it does, but LLMs are limited to their training data. Often that data was collected months, if not years ago (known as Recency Bias). Don't limit yourself to what the LLM "knows".

- **Emotional Nuance:** AI may struggle to understand or capture the emotional depth and subtleties required for certain creative endeavors. Don't be surprised when AI can't deliver the gut punch emotional reaction you want to summon in your audience. No AI system would come up with the Sarah McLachlan Animal Cruelty video (get ready for tears and adopting a few puppies after watching it). Or something as fun and smile worthy as the kid who loves corn on YouTube. Work with LLMs to generate ideas, mix those with your humanity to make the LLM's ideas impactful.

With these considerations in mind, you can wield the power of AI to break free from the chains of the creative block. Prepare to unleash your inner creativity beast, as you embark on a journey of imagination and innovation!

THE END OF THE BEGINNING

Once I was told, don't tell the war's story, tell the soldier's story and the war will be understood. We're not at war with AI, but we are in conflict with the future of work. The world is changing as fear and features of AI explode in every aspect of life.

Algorithms are already directing more of your life than you realize. Whether it is the recommendation engine on Amazon encouraging you to buy a toaster or Pinterest suggesting you follow a topic, algorithms are the true influencers of the internet age. As the cost to implement AI drops, every business can harness the power of AI in every team. Will your team be ready? Will you?

I hope this book helped you see yourself in the Age of AI. You can choose to be a leader or not. You can take charge of the AI revolution or be an engaged participant. The future of work isn't only for the disciples of Hustle Culture and 24/7 work. We need you too! People who just want to do their job and go home.

Normal is okay. AI can be normal.

I have one more exercise for you to do in this book. This one is my parting hope for you. Now is the time to truly learn.

EXERCISE: GROW ANOTHER

This exercise is an opportunity for you to step into the role of an educator and, in doing so, deepen your own understanding of AI. I encourage you to reach out to a local college, high school, mentoring program, youth group, whatever is in your area, and offer to run a workshop or seminar on AI. When you teach a topic, you truly learn it. I have always found that giving talks on a topic, teaching classes, leading conversations, has helped me transform information floating in my head into knowledge ready to be applied to a problem.

Here are four ideas you could use to spark your AI lesson:

Skill 1: Understanding Algorithms

The primary focus of this session should be on 'Understanding Algorithms'. Use real-life examples, like how Amazon's recommendation engine works or how Instagram uses algorithms to suggest connections. The aim is to ensure students grasp the practical implications of these algorithms in daily life. If you don't know how these things work, look up "How-To Build a Recommendation Engine" on YouTube and use what you learn with the content of this book to help people grow their AI skills. Remember, you don't have to write python code to be AI Literate.

Skill 2: Engaging with AI

This next session is focused on 'Engaging with AI'. Discuss how AI is becoming a part of every aspect of our lives and businesses. Explain how every team in a business can harness the power of AI, and the impact AI will have on their future work.

Skill 3: Understanding your work

Focus on helping people understand what they do at work every day. Explain how process analysis and taskification can help them understand their work, decompose the work into its parts, and objectively measure progress.

Skill 4: Problem Articulation

Teach students how to articulate problems they find in their work, and break them down into smaller, more manageable parts. Explain setting goals, defining success metrics, as well as finding existing solutions to similar problems online.

Remember, the best way to learn is to teach. As you prepare for and deliver these sessions, you will find your own understanding of AI deepening. And, who knows, you might just inspire the next generation of AI leaders.

WHAT To DO ON TUESDAY

If you haven't started your AI Journey yet, now's the time. Don't wait until Tuesday! Let's go now!

It is time to combine your magic with the power of AI. A normal future awaits!

ACKNOWLEDGMENTS

There is a lot that goes into a book. For this one, it was teaching, learning, experience, and the wonderful support of my family.

As always, the first thank you goes to Maria for helping make the space in our world for my writing.

Second, as always are my kids. This book was written for them and the hopes that they'll have a career in the world of AI that is as normal as they want it to be.

Thank you to my tech-checker Dr. Chris Kulp. He's a brilliant professor, machine learning wizard, and above all amazing brother. He has written some excellent textbooks and sci-fi fiction. Check him out on Amazon or your bookstore of choice.

Thank you to my students who asked the questions I wanted to answer in this book. Learning is such a critical thing for the future we're building and I hope you all keep to your journey.

And finally, thank you dear reader for going on this journey with me. I hope you found this book actionable, realistic and inspiring. Your future, and everyone else's, awaits, let's make it awesome.

ABOUT THE AUTHOR

Tim Kulp is a creative, techno-nerd, who loves AI and Mixed Reality technologies. He's the Chief Innovation Officer at technology consulting firm, Mind Over Machines in Baltimore, MD. When not playing with technology he is writing horror stories or enjoying time with his family. Tim has one friendly dog and one vengeful cat.

Follow Tim on his website: https://timkulp.com or LinkedIn at https://www.linkedin.com/in/timkulp.

Made in the USA
Las Vegas, NV
21 November 2023

81298199R00134